HIGH PERFORMANCE
MATERIALS AND DEVICES
FOR HIGH-SPEED
ELECTRONIC SYSTEMS

SELECTED TOPICS IN ELECTRONICS AND SYSTEMS

Editor-in-Chief: **M. S. Shur**

**The complete list of the published volumes in the series can be found at
https://www.worldscientific.com/series/stes

Selected Topics in Electronics and Systems – Vol. 61

HIGH PERFORMANCE MATERIALS AND DEVICES FOR HIGH-SPEED ELECTRONIC SYSTEMS

Editors

F. Jain
University of Connecticut, USA

C. Broadbridge
Southern Connecticut State University, USA

H. Tang
Yale University, USA

M. Gherasimova
University of Bridgeport, USA

W|e World Scientific

NEW JERSEY · LONDON · SINGAPORE · BEIJING · SHANGHAI · HONG KONG · TAIPEI · CHENNAI

Published by

World Scientific Publishing Co. Pte. Ltd.

5 Toh Tuck Link, Singapore 596224

USA office: 27 Warren Street, Suite 401-402, Hackensack, NJ 07601

UK office: 57 Shelton Street, Covent Garden, London WC2H 9HE

British Library Cataloguing-in-Publication Data

A catalogue record for this book is available from the British Library.

Selected Topics in Electronics and Systems — Vol. 61
HIGH PERFORMANCE MATERIALS AND DEVICES FOR HIGH-SPEED ELECTRONIC SYSTEMS

ISBN 978-981-3276-29-1

For any available supplementary material, please visit
https://www.worldscientific.com/worldscibooks/10.1142/11156#t=suppl

Desk Editor: Tay Yu Shan

Preface

In this review volume, we have invited experts working on micro/nano-electronics and optoelectronics/nano-photonics to report on state-the-art research and development in nanocomposites and optical electronics. In here, we have also expanded the scope to include contributions from the area of 3D printing and other emerging technologies:

- Enabling materials research papers include modeling of strain and misfit dislocation densities, microwave absorption characteristics of nanocomposites, and X-ray diffraction studies.
- Novel devices papers include quantum dot nonvolatile random access memory, spatial wavefunction switching (SWS), field effect transistors (FETs) and circuits, and quantum dot channel (QDC) FETs.
- The use of electrophoretic deposition technique to fill the deep silicon trenches by ^{10}B nanoparticles, which is used for thermal boron neutron detection, is the topic of one paper.
- In the area of emerging technologies, one paper presents additive manufacturing techniques to develop wireless modules, fully flexible energy autonomous body area network for autonomous sensing applications using 3D and inkjet printing techniques. Integration of inkjet and 3D printing for the realization of efficient mm-wave 3D interconnects up to 60GHz is also discussed.
- Optical modulators using quantum dot superlattice is the topic of one paper.
- Another paper presents experimental demonstration of a mode locked fiber ring laser with the implementation of a photonic crystal fiber (PCF) to generate pulse train at high speed.

In summary, papers presented in this volume involve various aspects of high performance materials and devices for implementing High-Speed Electronic systems.

Editors:

F. Jain (University of Connecticut)

C. Broadbridge (Southern Connecticut State University)

H. Tang (Yale University)

M. Gherasimova (University of Bridgeport)

Contents

Characterization of Ge Quantum Dot Optical Waveguides for High Speed Optical Modulators

Jan Amir Khan

Electrical Engineering, University of Connecticut, 10 Hillside Rd,
Storrs, CT, 06269 USA
Jan.Khan@uconn.edu

Evan Heller

Synopsys, 400 Executive Blvd, Suite 101,
Ossining, NY, 10562 USA

Faquir Jain[*]

Electrical Engineering, University of Connecticut, 10 Hillside Rd,
Storrs, CT, 06269 USA
fcj@engr.uconn.edu

Quantum Dot (QD) Optical Modulators can provide high speed modulation in low cost indirect bandgap materials. Si based optical modulators can be realized with the inclusion of self-assembled Ge QDs to provide low cost, high speed CMOS compatible optical devices. In this paper, we present the optical characterization of a novel Ge-QD Si-SiO$_2$ based waveguide for use in as an optical modulator. Optical performance figures of merit are investigated including insertion loss (IL) measurements, and Wavelength dependent loss (WDL). We present a multimode waveguide fabricated with conventional CMOS processing. The waveguide provides 4.43dB/cm loss and individual discrete absorption regimes corresponding to the unique minibands produced by superlattice properties of the self-assembled Ge QDs in the IR regime. Absorption properties of the Ge QDs are demonstrated and verified against the QD superlattice bandgap model. Analysis and simulation is presented to qualitatively compare the QD bandgap energies with the reported optical properties. The QD functionalized structure demonstrates the fundamental optical principles of a QD waveguide, setting the foundation for a active modulation testing of this QD based optical modulator.

Keywords: Quantum dot; quantum dot optical modulator; modulator; Si waveguide.

1. Introduction

Quantum Dots take advantage of discrete finite energy levels to provide increased electro-refraction (ΔN) and electro-absorption (ΔA) characteristics when compared to multiple quantum well and wire devices. Modeling and experimentation shows the advantage of the introduction of Ge quantum dot devices in indirect bandgap materials to enable high

---

[*]Corresponding author.

electro-refraction and absorption characteristics in indirect bandgap semiconductor. Quantum dot modulators will allow low voltage, high speed switching in comparison to conventional modulators. The realization of optical modulation utilizing QDs is realized first with basic waveguiding principles which are directly dependent on the material system, waveguide design, fabrication methods, and optical spectral parameters of the QD superlattice structure which forms a modified bandgap from bulk materials. Si/SiO$_2$ ridge waveguides offer large index of refraction delta of 40% with total internal reflection at large incident angle 60% [5]. The ability to properly waveguide in this low cost waveguiding medium sets the stage for the QD functionalization of the waveguide. In this paper, we will present a straight ridge optical waveguide with self-assembled Germanium core-Silicon Oxide cladded quantum dots applied to the waveguiding structure. The optical characteristics of the unmodulated device is presented. Absorption data is presented and compared with the band structure of the self-assembled QD layer.

2. Waveguide Design and Fabrication

A multimode ridge-based waveguide was simulated via RSOFT Beamprop software to ensure that the incident light propagates with relatively low loss over the modulator length. The design of the optical waveguide is engineered based on the optical wavelength, index different between core and clad, waveguide length, QD dot density and manufacturability limitations. The structure was simplified to allow ease of free space optical testing.

The QD layer is applied to the top of the optical Si ridge structure and then cladded with oxide. The QD superlattice layer is only 16nm thick and therefore the evanescent tail of the incident light wavefunction is overlapped into the QD layer. During testing this device acts like a Slab Coupled Optical Waveguide (SCOW) and the interaction with the QDs occurs on the top of the waveguide. The ridge structure was developed based on a 4.3um Mode Field Diameter (MFD) based fiber launch shown in Fig. 1(b). The waveguide presented in this paper features a thin ridge waveguide having a 0.4um height and 7um width.

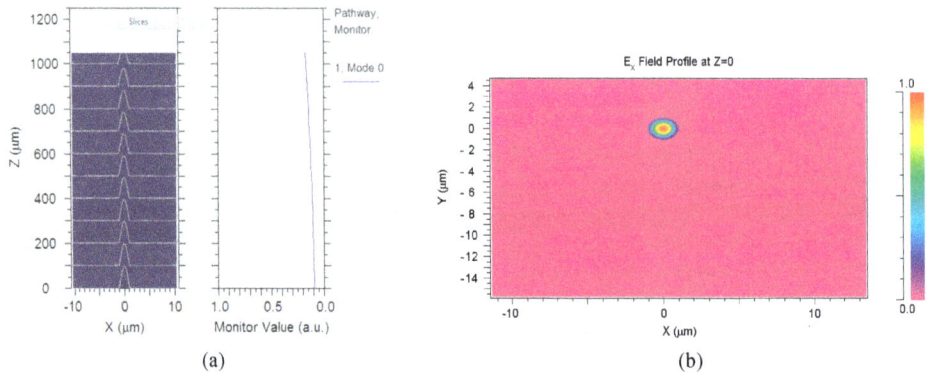

Fig. 1. (a) Optical waveguide propagation along Z axis. (b) Fiber launch profile.

The fabrication of a QD enhanced Si waveguide utilizes standard conventional cleaning, etching and lithography processing techniques while incorporating a novel liquid phase self-assembly process for Ge-O2QD deposition.

First a <100> P-type Si wafer is cleaned and subjected to Boron diffusion to increase the carrier concentration to $10^{17}cm^{-3}$. This provides proper charge for selective area self-assembly of Ge-GeO$_x$ cladded QDs later in the process. The handle of the SOI wafer is thinned from 625um to ~150um to provide easy cleaving for device testing via a fine lapping process. A 700A high quality pattern oxide is then grown on the device side of the wafer via dry oxide deposition. Utilizing a ridge pattern mask with the appropriate ridge width (7um), the non-ridge portion of the device is exposed. The surrounding oxide is etched via a buffer oxide etch to remove the oxide surrounding the ridge. Following the oxide removal, the Si layer is etched to create the ridge of the waveguide in the 5um Si device layer. A 13 second etching time was sufficient to remove ~300nm of Si. The removal of the pattern oxide is followed by growth of a thick 2200A cladding oxide. Photolithography is conducted once again to open the top of the ridge and remove the oxide layer via BOE.

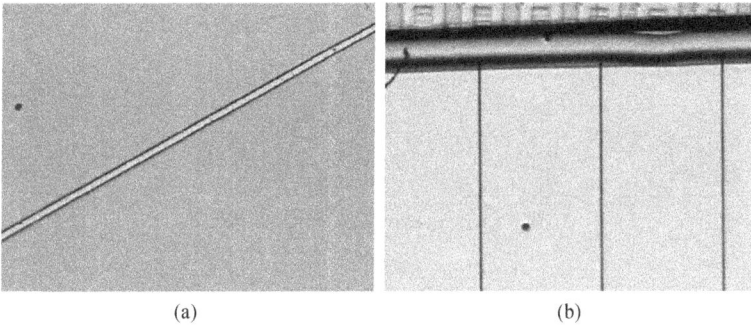

(a)   (b)

Fig. 2. (a) QD waveguide ridge top view. (b) Top view waveguide array.

Once the photoresist was removed, the device was processed for self-assembled QD deposition on the ridge structure. Self-assembly of Ge-GeO$_x$ QDs was processed according to previously developed colloidal synthesis techniques developed at the University of Connecticut Nanotechnology Laboratories. Once liquid self-assembly of the QD layer was performed, a top cladding layer of SiO$_2$ was applied. Multiple ridge waveguides were patterned on one sample as seen in Fig. 2(b).

3. Testing and Characterization

3.1. *Insertion Loss*

Insertion loss measurements were conducted across the full length of the waveguide via free space end-face coupling. The input light consists of 980nm SLD source with Full Width Half Maximum (FWHM) of 35nm and peak power of 16dBm. A custom tapered lensed fiber was utilized as the launch with mode field diameter (MFD) of 2.5um. This ensures proper coupling into the end face of the waveguide. Alignment of the input fiber to the height required submicron vertical alignment to provide small overlap between the ridge-device layer of Si and the small QD layer on top. Retaining high coupling efficiencies is imperative for proper QD waveguide coupling. The QD layer is approximately 40nm thick and therefore small wave overlap is essential for optical characterization. The output measurement fiber is a lensed fiber with 5um MFD spot size and carefully aligned to the top portion of the ridge waveguide to couple only the top part of the mode field. The power output of the SLD was increased and measurements were taken at both the output end and the input end of the launch fiber. The insertion loss vs drive current is plotted below.

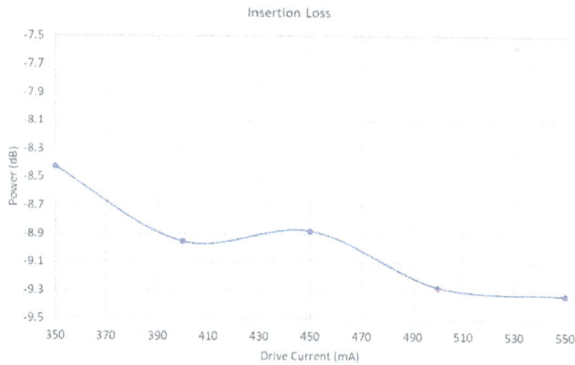

Fig. 3. Insertion Loss vs. Drive current.

Taking the intrinsic mismatched mode field diameters of the launch fiber and waveguide (6.54 dB) the estimated loss was -4.43 dB/cm. Higher than other Si-SiO$_2$ waveguides at 1550nm [3]. The high optical loss is primarily due to high sidewall roughness which has previously been shown can provide very high losses (-60dB/cm) in small core waveguides (0.4um x 0.2um waveguides) [5]. Concurrent measurements for optical absorption utilized partially coupled light from the top of the waveguide. This offset coupling was done to couple light propagating in the top portion of the ridge which featured self-assembled Ge QDs. This partial coupling increased the insertion losses to over -40dB which was expected. The partial coupling was imperative to demonstrate the functionality of the QD layer. In practical applications this will have detrimental effects in interconnect coupling for on-chip applications.

3.2. *Optical Absorption*

Optical absorption measurements were taken from 900nm up to 1600nm utilizing various broadband optical sources and a Yokogawa AQ6370 optical spectrum analyzer with 0.1nm resolution. Analysis of the optical absorption via wavelength dependent loss measurements were carried out from 920nm to 1060nm where the effective bandgap absorption coefficients demonstrate absorption proportional to the superlattice discrete bandgap energy levels. Previously, Germanium QDs superlattice structure has been modeled and simulated with the QD core and cladding effective bandgap values. The high confinement of a single QD presents an increased in the bandgap according to the following equation:

$$E_g \text{QD} = E_g^{bulk} + \frac{\hbar^2 \pi^2}{2R^2} \left(\frac{1}{m_e} + \frac{1}{m_h} \right)$$

The mass of the excited electron and holes are:

Ge QD: $m_h = 0.28$, $m_e = 0.08$

GeO$_x$: $m_h = 0.16$, $m_e = 0.16$

The QD superlattice provides additional increase in the effective bandgap which results in lower (<1um) wavelength absorption peaks. Single QD layer effective bandgap absorption and refractive index change has been demonstrated via modeling and simulation and is shown below.

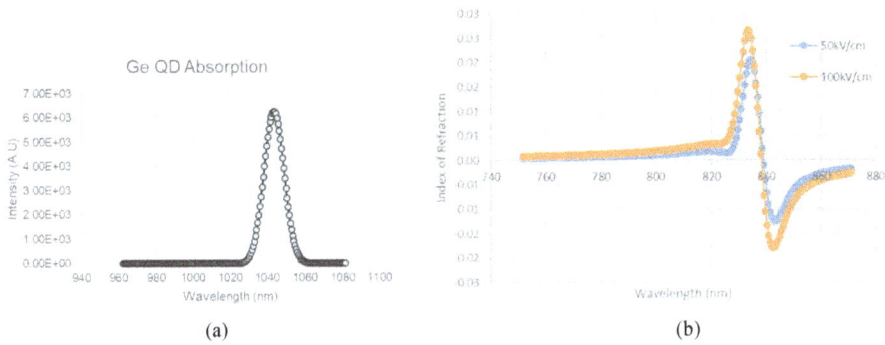

(a)   (b)

Fig. 4. (a) Absorption of single QD layer. (b) Refractive index change at distinct electric field.

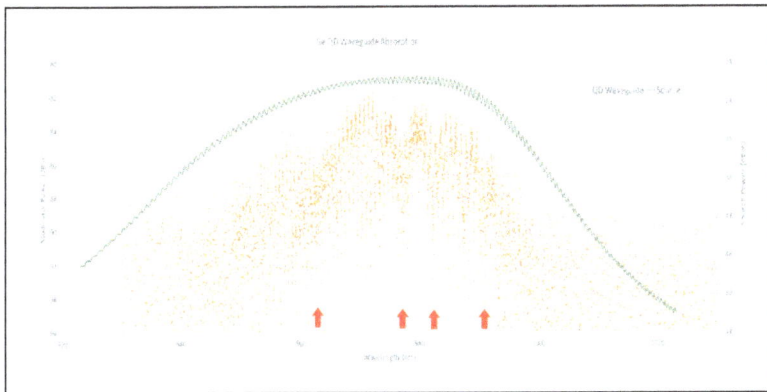

Fig. 5. Optical absorption measurement centered at 980nm.

Wavelength dependent loss due to absorption of the QD layer has been demonstrated.

The center absorption wavelengths are outlined in Table 1. It can be shown that four distinct absorption bands are discovered in the 960nm to 1um wavelength range. Each absorption band corresponds to a discrete energy bands in the QD superlattice.

Table 1.

Waveguide Ridge #3	Center Wavelength (nm)	FWHM (nm)	Corresponding Bandgap energy (eV)	Absorption (dBm)
Peak [1]	963.3	1.65	1.287241773	12.864
Peak [2]	977.35	2.61	1.268736891	14.19
Peak [3]	982.24	1.87	1.26242059	15.37
Peak [4]	991.04	1.68	1.251210849	13.09

Absorption peaks were recorded which coincide with the Ge QD bandgap energies. Where the Ge-QD core has a E_g of 0.67eV and GeO_x has a bandgap of 0.16eV. The effective bandgap corresponding to the superlattice layer is ~1.26eV as demonstrated in Table 1 above. The highest absorption band at 991nm demonstrates higher excitonic binding energies and corresponding lower wavelength optical absorption. This can be corelated to a reduced number of QD layers. The FWHM and direct wavelength of the absorption spectra demonstrate both QD dot uniformity and bandgap. A consistent 1.6nm FWHM shows exemplary QD uniformity the band structure with miniband separation of 18.5meV, 6.3meV and 11.2meV.

4. Conclusion

In this paper we reviewed the design, fabrication and basic optical characterization of a $Ge-GeO_x$ QDs in an $Si-SiO_2$ optical waveguide for use as a optical modulator. A multimode optical waveguide was constructed which features a SCOW structure with $Ge-GeO_x$ QDs self-assembled on the ridge structure. Optical characterization demonstrated -4.43db/cm insertion loss at 980nm which is respectable compared to previously explored losses in at 1550nm. For the first time, we demonstrate discrete optical absorption peaks which relate to the direct absorption of the QD superlattice structure. Large absorption peaks of 15dB were observed with the peak to peak variation in broadness being well maintained at 1.5nm FWHM indicating uniform QD size. The QD absorption peaks measured have shifted effective bandgap of ~0.02eV with the theoretical QD model values at 1um due to QD dot size. Further investigation is required under colder temps to more clearly demonstrate the excitonic absorption of the QD layer vs large thermal absorption which is prominent at room temperature.

References

1. Balanis, Constantine A. *Advanced Engineering Electromagnetics*. 2nd ed. Vol. 1. New York: Wiley, 1989. Print.

2. Grillot, F., L. Vivien, E. Cassan, and S. Laval. "Influence of Waveguide Geometry on Scattering Loss Effects in Submicron Strip Silicon-on-insulator Waveguides." *IET Optoelectron* 2.1 (2008): 1-5. *Ieeexplore*. Web. 8 Nov. 2012.
   <http://ieeexplore.ieee.org/stamp/stamp.jsp?arnumber=04455546>.
3. Shutler, Alisha. "Investigation of Propagation Loss of Passive Silicon Waveguides Using Two Different Etching Techniques." *NNIN REU Site: Nanotech* (2008): 114-15. *University of California, Santa Barbara*. Web. 8 Dec. 2012.
   <http://www.nnin.org/sites/default/files/files/2008NNINreuSHUTLER.pdf>.
4. Yamada, Hirohito, Tao Chu, Satomi Ishida, and Yasuhiko Arakawa. "Si Photonic Wire Waveguide Devices." *IEEE Journal of Selected Topics in Quantum Electronics*, 12.6 (2006): 1371-379. *IEEE*. Web. 8 Dec. 2012.
   <http://ieeexplore.ieee.org/xpls/abs_all.jsp?arnumber=4026596&tag=1>.
5. Yamada, Koji. "Silicon Photonic Wire Waveguides: Fundamentals and Applications." *Topics in Applied Physics*, 119 (2011): 1-29. Web. 8 Nov. 2012.
   <www.springer.com/cda/content/.../9783642105050-c1.pdf?...0-0>.
6. J. Khan, M. L.-Y. (2012). Quantum Confined Stark Effect in Silicon Quantum Dot Field Effect Structure. *Connecticut Microelectronics Optics Conference.*

Electrophoretic Deposition of ^{10}B Nano/Micro Particles in Deep Silicon Trenches for the Fabrication of Solid State Thermal Neutron Detectors

Machhindra Koirala[1], Jia Woei Wu[1], Adam Weltz[2], Rajendra Dahal[1], Yaron Danon[2], and Ishwara Bhat[1,*]

[1]*Department of Electrical, Computer, and Systems Engineering,*
Rensselaer Polytechnic Institute, Troy, NY, 12180, USA
[*]bhati@rpi.edu
[2]*Department of Mechanical, Aerospace, and Nuclear Engineering,*
Rensselaer Polytechnic Institute, Troy, NY, 12180, USA

We present a cost effective and scalable approach to fabricate solid state thermal neutron detectors. Electrophoretic deposition technique is used to fill deep silicon trenches with ^{10}B nanoparticles instead of conventional chemical vapor deposition process. Deep silicon trenches with width of 5-6 μm and depth of 60-65 μm were fabricated in a p-type Si (110) wafer using wet chemical etching method instead of DRIE method. These silicon trenches were converted into continuous p-n junction by the standard phosphorus diffusion process. ^{10}B micro/nano particle suspension in ethyl alcohol was used for electrophoretic deposition of particles in deep trenches and iodine was used to change the zeta potential of the particles. The measured effective boron nanoparticles density inside the trenches was estimated to be 0.7 gm cm^{-3}. Under the self-biased condition, the fabricated device showed the intrinsic thermal neutron detection efficiency of 20.9% for a 2.5 × 2.5 mm^2 device area.

Keywords: Neutron detectors; electrophoretic deposition; silicon deep trenches; anisotropic etching.

1. Introduction

Solid state thermal neutron detector devices are pursued as possible replacement for He-based detectors because these detectors minimize many disadvantages of He-based detectors such as device bulkiness, high voltage requirement *etc.* [1-2]. Silicon devices themselves do not interact with thermal neutrons and hence a converter material such as LiF, ^{10}B, ^{157}Gd *etc.* has to be incorporated. These converter materials produce daughter particles that are charged and can be detected by corresponding silicon device [3-4]. A typical device involves fabrication of silicon pn junction and coating it by a thin boron. To get high efficiency, microstructured silicon pn junction devices are being used. This involves filling of silicon microstructure with ^{10}B [5] and the alpha particles produced during the interaction between thermal neutron and ^{10}B can reach to p-n junction. These

---

[*]Corresponding author.

alpha particles create electron-hole pairs in p-n junction. The electrons and holes are separated by the junction built-in potential and can be detected as an electrical signal. To optimize the neutron detection efficiency, the interaction of neutron with [10]B and charge collection by nearby p-n junction has to be optimized at the same time. To carry out such optimization, many micro-structured silicon devices have been proposed such as honeycomb structures, pillar structures, silicon trenches structures [6-8]. Among these microstructured silicon p-n junction for thermal neutron detection, honeycomb and pillar structures were fabricated by using deep reactive ion etching (DRIE) method [9] and have been reported previously.

DRIE method is expensive and may not be suitable for cost effective mass production. To overcome this situation, we proposed to modify the device fabrication method by replacing DRIE method by wet chemical etching method. Trenched silicon structure can be fabricated using anisotropic wet chemical etching of (110) silicon wafers following the technique developed in 1960's [10-11]. Previously, we have used low pressure chemical vapor deposition method (LPCVD) to fill the deep holes in silicon p-n junction to fabricate devices. However, for a trench structure, LPCVD deposition of boron film encounters many problems. One of the main problems is the stress introduced by the boron filling process. To overcome this stress problem caused by LPCVD, the filling of boron using nanoparticles was proposed. However, the nanoparticle filling results in less dense boron and this necessitates deeper trenches and thicker trench width. So, we explored electro-phoretic deposition method for filling boron nano/micro particles in high aspect ratio silicon trenches and this paper addresses some results on the boron filling of trenches using boron nanoparticles. Device results on p-type silicon wafer and the results so obtained are also presented here for the first time.

Electrophoretic deposition is a colloidal processing technique in which thin film can be coated in a structured surface. The deposition process relies on the surface charge of the particles in a certain dispersive medium which determines the corresponding potential in its vicinity which introduces repulsive interaction among the particles. This potential is measured in terms of zeta potential which is defined as the potential at the share plane of the particle [12]. The dispersion medium having the absolute value of zeta potential higher than 30 mV is considered as a stable dispersion [13]. The zeta potential determines the rate of diffusion of a charged particle in the medium in an applied DC electric field. The absolute value of the zeta potential can be changed by changing the pH value of the dispersion medium. Addition of surfactant changes the charge balance condition and hence changes the zeta potential of the system. For example, the zeta potential of the particle in certain medium can be changed by adding iodine. Iodine produces H^+ ion in some organic solvents and changes the pH value of the liquid [14].

It has been reported that the electrophoretic deposition can be used to deposit thin film on to a structured surface with different spatial deposition selectivity. Generally, the pattering of the electrodes for electrophoretic deposition is done to control the materials deposition area and particles assembly [16]. Nadal *et al.*, have successfully demonstrated that the dielectric strip on the electrode can change the hydrodynamic flow of the electrolyte solution [17] and colloidal particles can be deposited in a featured surface. In a similar fashion, Zhang *et al.*, have demonstrated the driving of CdSe nanoparticles in a template with nanoscopic features [18]. The method of electrophoretic deposition in a featured surface can be easily implemented to drive the boron nanoparticles in deep trenches using the parallel plate electrode model.

The modification of boron filling method is favorable from the device fabrication prospective as well. One of the key problems with our previous version of solid state thermal neutron detector was that after filling the trenches with LPCVD boron, a window opening is required to make front contact. To open an ohmic contact window in a honeycomb structure, dry etching of boron was done using SF_6 gas. Due to poor selectivity between boron and silicon etching, there is always the risk of top layer of silicon getting etched through the p-n junction and the device getting shorted. Using electrophoretic deposition technique, the silicon trenches can be filled at room temperature and the contact metallization on the device can be made even before boron nanoparticles filling. The boron nanoparticles at the contact surface can be simply wiped off after the deposition to make the contact.

This paper reports for the first time on the fabrication of the neutron detector in a p-type silicon wafer with boron filling carried out by the electrophoretic deposition process. The reason behind choosing a p-type wafer is that the p-region will be bulk of the wafer that absorbs the alpha particles generated by the boron converter material. Since the minority carriers are electrons, higher collection efficiency is possible since they have larger diffusion length compared to holes; they can travel much faster and can be collected by the thicker p-type region. In fact, in the silicon photovoltaic industry, 84% of the solar cells use p-type wafer as the absorber layer.

2. Experimental Procedure

2.1. *Fabrication of silicon trenches*

The schematic diagram of the fabricated device is presented in Fig. 1. A p-type (110) wafer with carrier concentration of ~2 x 10^{15} cm^{-3} is used as a starting wafer. Front side of the wafer is doped n-type by diffusing phosphorous. The diffusion is done at 875 °C for 10 minutes using $POCl_3$ as pre-dep followed by diffusion for 40 minutes. The phosphosilicate glass grown during phosphorus diffusion was etched away using buffered oxide etch (BOE) solution. Device isolations were done using the dry etching of silicon using SF_6 in oxygen plasma which separates the dies with unit area of 2.5 mm x 2.5 mm. The exposed mesa side walls were passivated using 1.5 um silicon dioxide. On the device area, silicon dioxide of thickness of 300 nm was deposited using plasma enhanced chemical vapor deposition (PECVD) method and the mask for trench etching was made using photolithography followed by plasma etching of oxide by CHF_3 gas in oxygen plasma. High aspect ratio silicon trenches were made using wet etching of silicon using TMAH at 100 °C for 40 minutes. From the wet etching of silicon, we obtained the silicon trenches with width of 5-6 µm and depth of 60-65 µm. The side wall of the silicon trenches is

2-3 μm. After trench etching, the sample was RCA cleaned and a continuous p-n junction was made using $POCl_3$ diffusion at 825 °C with 10 minutes pre-dep and 30 minutes to diffuse phosphorus. The back side of the wafer is coated with Al using sputtering and the Al coated wafer is annealed at 400 °C for 2 minute using rapid thermal annealing (RTA). After the fabrication of the sample, the trenches were filled by enriched boron nanoparticles using electrophoretic deposition process.

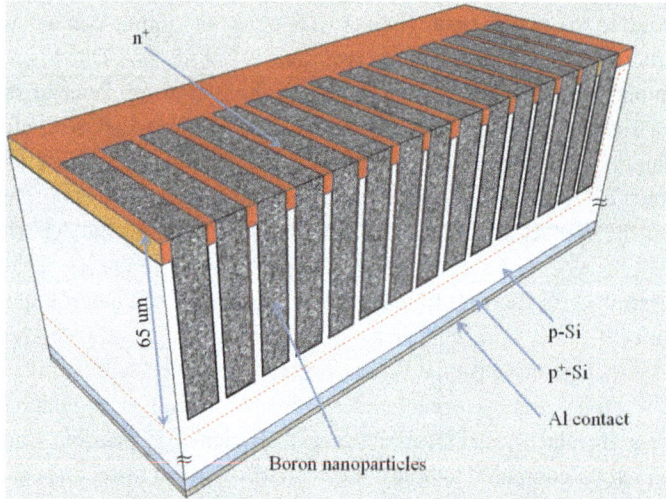

Fig. 1. Schematic cross-section of silicon trenched thermal neutron detector.

2.2. *Analysis of enriched boron micro/nano particle suspension*

The trenches are filled by boron using suspension of boron nanoparticles in ethyl alcohol. We have analyzed the surface charge behavior of ^{10}B micro/nano particles in ethyl solution. To enhance the surface charge density, we have used iodine as a surfactant. We prepared 3 different sets of boron nanoparticles. For the first one, we had used 0.1 mg of boron nanoparticles in 20 ml of ethyl alcohol. In the second suspension, 4 mg of iodine was dissolved in 20 ml of ethyl alcohol and 0.1 mg of boron nanoparticles was added. In the third suspension, 8 mg of iodine was dissolved in ethyl alcohol and 0.1 mg of boron nanoparticles was added. All of these three suspensions were sonicated separately and the zeta potential of these suspensions were measured using Zetasizer (Nicomp-PSS). The results of the zeta potential measurement are presented in Fig. 2. The zeta potential increased from 6 mV without iodine to 11 mV with the addition of 8 mg of iodine. Hence, for boron micro/nano particles, the zeta potential, which controls the electrophoretic deposition rate can be changed using iodine as a surfactant.

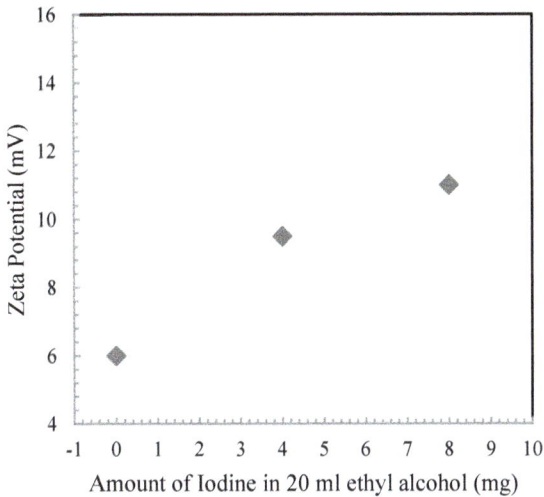

Fig. 2. Zeta potential of boron (^{10}B) micro/nano particle in ethyl alcohol with added iodine as a surfactant.

2.3. *Silicon trench filling*

To make the dispersion system for electrophoretic deposition, 40 mg of iodine was added and dissolved in 100 ml of ethyl alcohol (similar to adding 8 mg of iodine in 20 ml ethyl alcohol for the zeta potential measured suspension). Then, 1 gm boron nanoparticles were added to that ethyl alcohol and ultra-sonicated for 10 minutes. The schematic diagram for the electrophoretic deposition is shown in Fig. 3. As shown in the figure, silicon device is

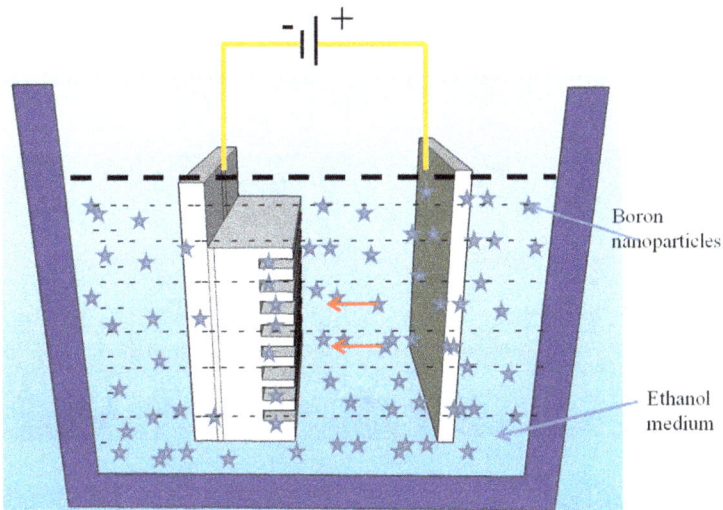

Fig. 3. Schematic diagram of experimental set up for electrophoretic deposition used for filling enriched boron micro/nano particles in silicon trenches.

attached in the cathode side. In this experimental configuration, the trajectory of the ^{10}B particles is perpendicular to the plate and is parallel towards the depth of the trench. After the device filling, the top part of the sample was wiped out to clean any over deposition. One part of the sample is cleaved and scanning electron microscope (SEM) image was taken. For the other piece of the sample, sputter coating of Ti/Al was done on the front part of the sample to make metallic contact. The sample was wire bonded using silver epoxy and I-V, C-V characterization and neutron detection efficiency was measured.

3.   Results and Discussions

In this section, we will discuss the device characterization involving boron nanoparticles filling, electronic properties of the fabricated device and the device efficiency for neutron detection.

3.1.   *Boron nano/micro particles filling fraction*

By weighing the silicon sample before and after nanoparticles filling, we could estimate the density of the filled nanoparticles. The volume of the trench is estimated by using pitch of the trench devices, width of silicon wall, width of trenches, and the depth of the trenches. The filling density of nanoparticles inside the trench was estimated to be 0.70 gm cm^{-3}.

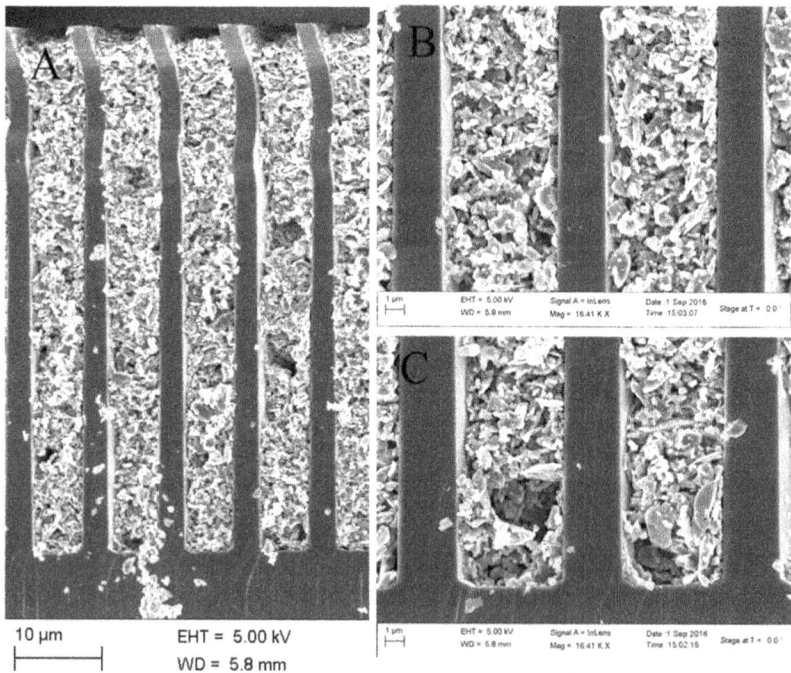

Fig. 4. SEM image of the cross-section ^{10}B filled silicon trenches. (A) SEM image of the filled sample. (B) Top part of the filled trench device. (C) Bottom part of the filled trench devices.

We obtained lower density for the nanoparticle filled trench devices compared to our previous LPCVD boron filled devices [19] which was 1.8 gm cm^{-3}. However, the depth of the holes in honeycomb structure were 45 μm while the depth of the silicon trenches is 60 μm giving larger space to fill boron. After boron nanoparticles were filled using electrophoretic deposition, the sample was cleaved and the images of the filled trench were taken using scanning electron microscope (SEM, SUPRA). The SEM images of the cross-section of samples are presented in Fig. 4. From the SEM images, we have seen that the micro/nano particles were filled up to the bottom of the trenches.

3.2. *Leakage Current*

Leakage current of the fabricated neutron detector p-n junction is important since it indicates the noise level of the detector. The leakage current was measured and the results are presented in Fig. 5. While calculating the device current density, we considered the top area of the device as the device active area which is 0.0625 cm^{-2} for a 2.5 x 2.5 mm^2 die. In fact, this is not the true surface area of the p-n junction because of the trench profile of the device. The wall of the device is continuous p-n junction which has estimated surface area of the silicon wall is of the order of 0.72 cm^2, so the effective leakage current density is one order of magnitude less than the reported value. Compared to our previous detector, the leakage current density in this device is high [15].

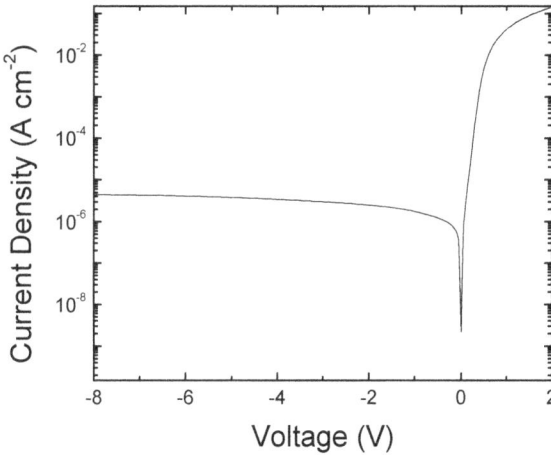

Fig. 5. I-V properties of 2.5 mm x 2.5 mm silicon neutron detector.

3.3. *Capacitance*

One of the key parameters that affects the performance of neutron detector is the capacitance of the device. If the capacitance of the sample is high, the detection limit of the device goes down. Figure 6 shows the capacitance as a function of reverse biased voltage. The capacitance of 2.5 x 2.5 mm^2 device at 0 V is 7 nF which sharply decreases to 1 nF at 1V reverse biased voltage. To understand the capacitance behavior, we fabricated and measured the capacitance for a planner device of the same wafer and estimate the carrier concentration using Mott-Schottky relation as shown in Eq. (1).

$$\frac{1}{C^2} = \frac{2}{q\varepsilon_{si}} \frac{1}{N_d} (V_0 + V_R) \qquad (1)$$

where C is the capacitance per unit area, q is electronic charge, ε_{si} is permeability of silicon, N_d is the doping concentration of the wafer, V_0 is the barrier potential of silicon and V_R is the reversed biased voltage.

From Eq. (1), the carrier concentration of the wafer is calculated to be 1.7×10^{15} cm^{-3}. Using the calculated carrier concentration, the depletion width of the p-n junction is estimated to be 750 nm. The thickness of the trench walls are 2-3 um which is more than double of the depletion width. Because of the wider trench width, the devices are not fully depleted and the side wall of the trenches also contributes to capacitance. From the geometrical structure of the trenches, the surface area of the trenches is 0.72 cm^2 which is much higher compared to the planner area of the device which is 0.0625 cm^2. For the undepleted condition, the effective area of the capacitor becomes 0.72 cm^2 while for the fully-depleting-condition the effective area of the capacitor become 0.0625 cm^2. This is the main reason why the capacitance of the device is very high compared to planner diode of same area at zero biased condition. Using the reverse biased condition, we have calculated the capacitance for these two device areas and the results along with the measured capacitance data are presented in Fig. 6. From the capacitance plot, we can see that, a reversed biased voltage of 1 V is required for the device to get fully-depleted

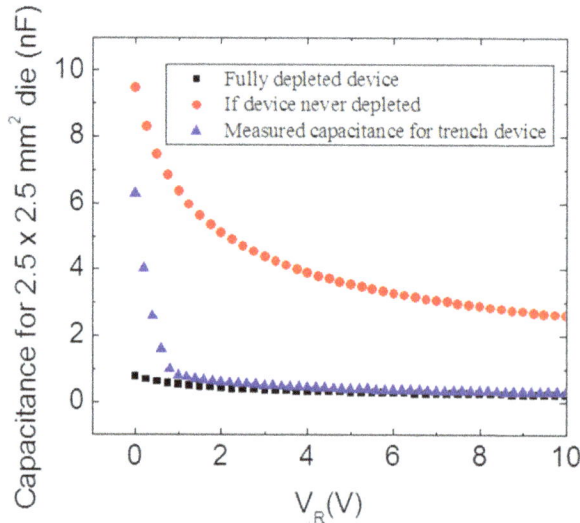

Fig. 6. Capacitance measured at different reverse biased voltage measured at 100 KHz. The capacitance has been calculated using the trenched structure with fully depletion and without depletion condition.

condition. From the capacitance point of view, there are couple of ways to improve the device to reduce the noise, one of them is to make the thinner trench with thickness less than 1.5 µm (twice of the depletion width). Since, the wall with thinner width will be fragile and will be very hard for further processing such as nanoparticles filling, another way to improve the device performance is to use higher resistivity wafer. In such case the depletion width of the p-n junction will be larger and the fully depleted trenches can be obtained.

3.4. *Neutron Counts*

The efficiency of thermal neutron detector was done using calibrated fission neutron source of ^{252}Cf which is moderated by high density polyethylene housing of 61 x 61 x 40 cm^3. The neutron flux is calibrated and found that it has the flux of 298 n cm^{-2} s at the distance of 8 cm from the source. We measured the neutron count for our devices at different conditions (i) at the lead housing when there is no neutron source in the room, (ii) at 8 cm away from the neutron source, (iii) with 2 mm thick Cd sheet in between neutron source and the detector, and (iv) with ^{60}Co source to measure the gamma response. The result of the neutron response measurement is presented in Fig. 7. When there is no source and the device is in the lead brick housing, the device has the noise level of 600 KeV. Then the device is measured using ^{252}Cf source which gives the count for thermal as well as fast neutron. Later, a Cd shielding of 2 mm width has been used to shield the thermal neutron and only high energy neutrons were detected. We subtracted the high energy neutron counts to obtain the thermal neutron counts. We obtained the thermal neutron counts of

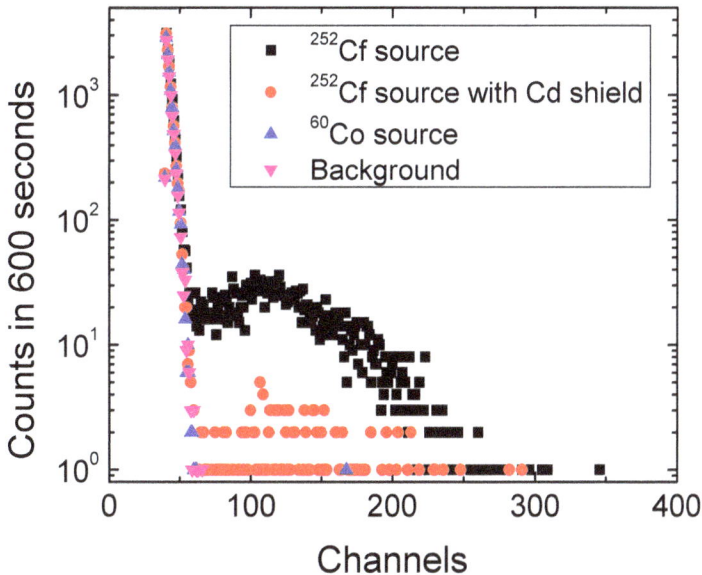

Fig. 7. Pulsed height spectrum for the trenched structured neutron detector filled with boron micro/nano particles.

2333 in 600 seconds for the incoming neutron flux of 298 n cm^{-2} s. Figure 7 shows the count rate of thermal neutrons at different energy ranges. From the data, we have found the detection efficiency of our device is to be 20.9%.

4. Detector Cost Reduction

Mass production is needed to diversify the use of solid state neutron detector over the ^{3}He gas based neutron detector. To reduce cost, our ^{10}B filled silicon trench based thermal neutron detectors is a viable alternative. We made key changes in the device processing methodology to reduce the device cost. One of the methods used in our detector is the wet chemistry method for trench etching and is an easy technique compared to DRIE method for making silicon trenches. The use of electrophoretic deposition method is also a cost effective method over LPCVD. The low pressure CVD method involves the use of expensive enriched diborane gas. The filling using electrophoretic deposition can use boron nanoparticles and that can be done by using some home-made customized tools.

5. Conclusions

In conclusion, we have implemented the wet etching method for trench etching and electrophoretic deposition for filling the silicon trenches by ^{10}B nanoparticles. The loosely filled nanoparticles has the density of 0.7 gm cm^{-3} in silicon trenches of 5-6 μm width and 60-65 μm depth. We have successfully demonstrated that the EPD filling method can fill to the bottom of the high aspect ratio silicon trenches. Following the process mentioned earlier, we obtained the neutron detection efficiency of 20.9% for solid state thermal neutron detectors. This is the first publication of neutron detector fabrication in p-type silicon using electrophoretic deposition process for boron filling.

Acknowledgement

The authors would like to thank the support staff of the Rensselaer Polytechnic Institute Micro-Nano-Clean-Room. This work was financially supported by USDHS/DNDO under grand number ECCS-1348269 and 2013-DN-077-ER0001.

References

1. GAO (U. S. Government Accountability Office), *Technology assessment: Neutron detectors: alternatives to using Hellium-3*, 2011, GAO-11-753.
2. J. F. C. A. Veloso, F. D. Amaro, J. M. F. dos Santros, J. A. Mir, G. E. Derbyshire, R. Stephenson, N. J. Rhodes and E. M. Schooneveld, Application of the microhole and strip plate detector for neutron detection, *IEEE Transection on Nuclear Science*, **51**, 2104 (2004).
3. D. S. McGregor, M. D. Hammig, Y.–H. Yang, H. K. Gersch, R. T. Klann, Design consideration for thin film coated semiconductor thermal neutron detectors-I: basics regarding alpha particles emitting neutron reactive films, *Nucl. Instrum. Methods Phys. Res. A*, **500**, 272 (2002).
4. D. S. McGregor, W. J. McNeil, S. L. Bellinger, T. C. Unruh, J. K. Shultis, Microstructured semiconductor neutron detectors, *Nucl. Instrum. Methods Phys. Res. A*, **608**, 125 (2009).

5. F. Sarubbi, T. L. M. Scholtes, L. K. Nanver, Chemical vapor deposition of alpha-boron layers on silicon for controlled nanometer-deep p$^+$n junction formation, *J. Electron Mater.* **39**, 162 (2010).

6. N. LiCausi, J. Dingley, Y. Danon, J.–Q. Lu, I. B. Bhat, A novel solid-state self-powered neutron detector, *Proc. of SPIE*-**7079**, 707908 (2008).

7. R. J. Nikolic, Q. Shao, L. F. Voss, A. M. Conway, R. Radev. T. F. Want, M. Dar, N. Deo, C. L. Cheung, L Fabris, C. L. Britton and M. N. Ericson, Si pillar structured thermal neutron detectors: fabrication challenges and performance expectations, *Proc. of SPIE*-**8031**, 803109 (2011)

8. K.–C. Huang, R. Dahal, J. J.–Q. Lu, A. Weltz, Y. Danon, I. B. Bhat, Scalable large-area solid-state neutron detector with continuous p-n junction and extremely low leakage current, *Nucl. Instrum. Methods Phys. Res. A*, **763**, 260 (2014).

9. K. C. Huang, R. Dahal, J. J.–Q. Lu, Y. Danon, I. B. Bhat, Boron filling of deep holes for solid-state neutron detector applications, *Transection of American Society*, **106**, 105 (2012).

10. J. M. Crishal and A. O. Harrington, *Electrochem. Soc. Extended Abstract*, Los Angeles, CA, 1962, Abs. No 89.

11. D. B. Lee, Anisotropic etching of silicon, *J. Appl. Phys.*, **40**, 4569 (1969).

12. L. Bousse, S. Mostarshed, B. V. D. Shoot, N. F. de Rooij, P. Gimmel, W. Gopel, Zeta potential measurement of Ta$_2$O$_5$ and SiO$_2$ thin films, *J. Colloid Interface Sci.*, **147**, 22 (1991).

13. J. D. Clogston, A. K. Patri, Zeta potential measurement, *Methods Mol. Biol.*, **697**, 63 (2011).

14. L. Dusoulier, R. Cloots, B. Vertruyen, R. Moreno, O. Burgos-Montes, B. Ferrari, YBa$_2$Cu$_3$O$_{7-x}$ dispersion in iodine acetone for electrophoretic deposition: Surface charging mechanism in a halogenated organic media, *J. Eur. Ceram. Soc.*, **31**, 1075 (2011).

15. R. C. Bailey, K. J. Stevenson, J. T. Hupp, Assembly of micropatterned colloidal gold thin film via microtransfer molding and electrophoretic deposition, *Advanced Materials*, **12**, 1930 (2000).

16. F. Qian, A. J. Pascal, M. Bora, T. Y.–J. Han, S. Guo, S. S. Ly, M. A. Worsley, J. D. Kuntz, and T. Y. Olson, On-demand and local selective particle assembly via electrophoretic deposition for fabricating structure with particle-to-particle precision *Langmuir*, **31(12)**, 3563 (2015).

17. F. Nadal, F. Argoul, P. Kestener, B. Pouligny, C. Ybert, and A. Ajdari, Electrically induced flows in the vicinity of a dielectric stripe on a conducting plane, *Eur. Phys. J. E*, **9**, 387 (2002).

18. Q. Zhang, T. Xu, D. Butterfield, M. J. Misner, D. Y. Ryu, T. Emrick, and T. P. Russell, Controlled placement of CdSe nanoparticles in diblock copolymer template by electrophoretic deposition, *Nano Letts.*, **5(2)**, 357 (2005).

19. R. Dahal, K. C. Huang, J. Clinton, N. LiCausi, J.–Q. Lu, Y. Danon, I. B. Bhat, Self-powered micro-structured solid state neutron detector with very low leakage current and high efficiency, *Appl. Phys. Lett.*, **100**, 243507 (2012).

Dual Quantum Dot Superlattice

Barath Parthasarathy

Electrical and Computer Engineering at University of Connecticut, 371 Fairfield Way,
Storrs, Connecticut, 08889, USA
barath.parthasarathy@uconn.edu

Pial Mirdha

Global Foundries, 20070 NY-52, Hopewell Junction,
New York, 12533, USA
pial.mirdha@uconn.edu

Jun Kondo*, Faquir Jain[†]

Faquir Jain Group, University of Connecticut, 371 Fairfield Way,
Storrs, Connecticut, 08889, USA
*kondo@hartford.edu
[†]fcj@engr.uconn.edu

In this paper, we propose a structure using four layers of quantum dots on crystalline silicon. The quantum dots site-specifically self-assembled in the p-type material due to the electrostatic attraction. This quantum dot super lattice (QDSL) structure will be constructed using a mixed layer of Germanium (Ge) and Silicon (Si) dots. Atomic Force Microscopy results will show the accurate stack height formed from individual and multi stacked layers. This is the first novel characterization of 4 layers of 2 separate self assemblies. This was also applied to a quantum dot gate field effect transistor (QDG-FET).

Keywords: Quantum dots; quantum dot supper lattice; atomic force microscopy; quantum dot gate FETs; multi-state.

1. Introduction

Quantum dots (QDs) are unique materials with inner cores made of intrinsic silicon/ germanium and exterior cladded cores of oxide [1]. These particular QDs assemble in only two layers per application. This has been taken advantage of to produce novel devices Multi-Valued Logic (MVL) that operate in ternary and quaternary logic. In addition memory devices such as Flash Memory (NVM) based QDG-NVM [1] has been demonstrated and a two bit memory has been demonstrated.

[†]Corresponding author.

2. Quantum Dot Self-Assembly

When assembling cladded quantum dots (QDs), the dots will self-assemble only on the p-regions of the device. This occurs because the QD's outer cladding layer tends to be negatively charged, thus attracted towards the positive p-region. The user can then selectively deposit the dots over specific areas with rather minimal difficulty. The dots of silicon or germanium assemble with two layers during each self-assembly process. The process of how quantum dots are created is as follows.

A container with silicon or germanium powder is placed in a nitrogen environment for up to five hours. The container of powder is then ball milled for five hours, this process causes the powdered crystals to shrink from the micron range to nanometer ranges. Once ball milled down to smaller sized particles, the Si/Ge powder is combined to form a solution with benzoyl peroxide and ethyl alcohol. This helps to oxidize the powdered solution and ethyl alcohol to keep the solution suspended. The solution is then placed in a sonicator to further help the oxidation process and prevent the QD from amalgamating. The solution is then spun in a centrifuge to separate out the larger QDs, in the ~20 nm ranges, to smaller ones in the 10 nm range. This is then chemically etched down to get the desired dot sizes of ~6-8 nm for Silicon and ~3-4 nm for Germanium. Previous DLS measurements have shown the dot size to be very large prior to sonicating and centrifuging the solution [2]. The silicon sample where QDs are being applied is then placed in this colloidal solution for 5 minutes. Once completed, the samples are rinsed in methanol and annealed in an argon environment [2].

As previously mentioned each self assembly process can only deposit two layers because the self assembly process is a function of time and water within the solution [3-5]. 4 layers can be obtained by extending the process listed above. Cladding layer chemistry for concurrent assemblies is helped by the immersion of De-ionized (DI) water. The application of DI water helps form the –OH bonds on the cladding layer [1]. The sample is rinsed in DI water for 3 minutes after the first annealing has been completed and before allowing the 2nd self-assembly to begin. This allows for hydrolysis of the Si-O groups on the secondary layer providing enough absorption for possible third and fourth layers. Annealing it after performing a second self assembly will finalize the four dot layer structure. This method will be explored when trying to build a sample with four layers. Basic line samples have gone through this process and are surfaced profiled using atomic force microscopy (AFM).

To test dot size and correct distribution, the dots were first deposited on alternating p-type and n-type doped lines allowing for the surface profile to represent alternating regions with and without dots as the QDs only assemble on p-type surfaces. Figures 1 and 2 shows the AFM surface potential profiles of line patterned with two layers of Si and Ge dot formation, respectively. The higher stacked pillars represent the areas where dots have been assembled on the line sample. Using the AFM measurements, the size of the individual dot sizes can be characterized. The germanium dot size was measured to be ~4nm per layer, while the silicon dot size was measured to be ~6nm per layer. Since the

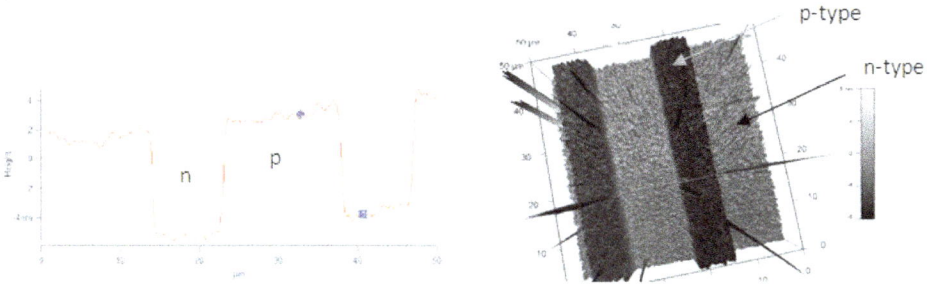

Fig. 1. Germanium AFM.

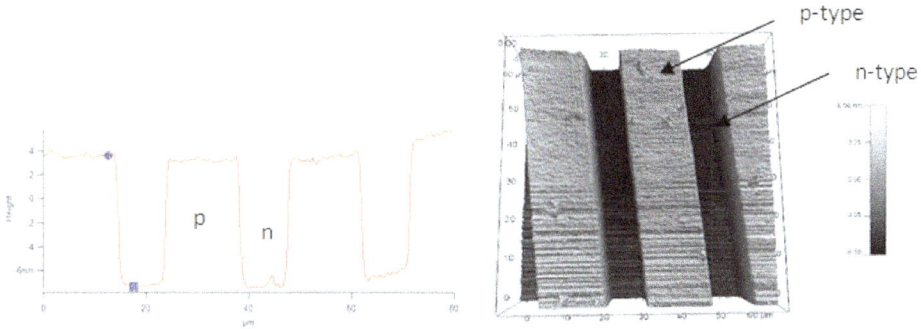

Fig. 2. Silicon AFM profile.

dots assemble in two layers, Figs. 1 and 2 show accurate height values when assembled on two different samples.

Subtracting the top and bottom of the square stack, helps determine an overall height attained from one self-assembly of silicon and germanium. The Fig. 1 germanium stack height average can be measured to be ~8nm, 4nm per dot. The Fig. 2 silicon stack height average is measured to be ~12nm, 6nm per dot. Both figures show confidence that the height produced by one self-assembly is accurate.

The three dimensional view shows mostly uniformity with some periods of noise. To get rid of the impact of the noise, a larger height area is averaged to overcome these extraneous points.

Being able to achieve samples with 4 layers of dual Si/Ge stacked dots is important step forward to extending QD functionality. Figures 3 and 4 show the surface profile for a QDSL made up of Silicon dots on bottom and Germanium dots on top. In Fig. 4, the red line across the 2-D profile represents the area that was averaged prior to obtaining Fig. 3. Averaging larger areas can help minimize the influence of noise. The total height average for the QDSL was ~18nm. This is accurate with the sum of individually self-assembled silicon (12) and germanium (6) samples that was also obtained from Figs. 1 and 2.

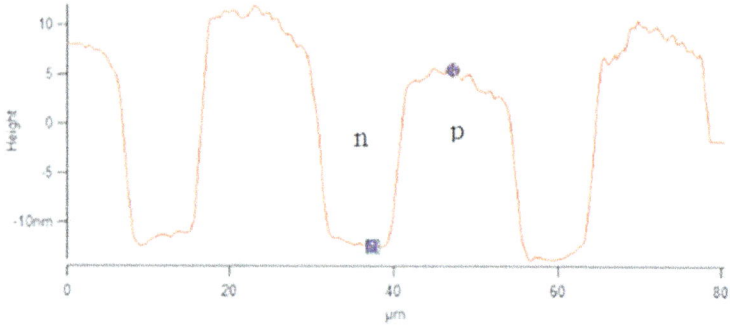

Fig. 3. Silicon and Germanium stacked profile.

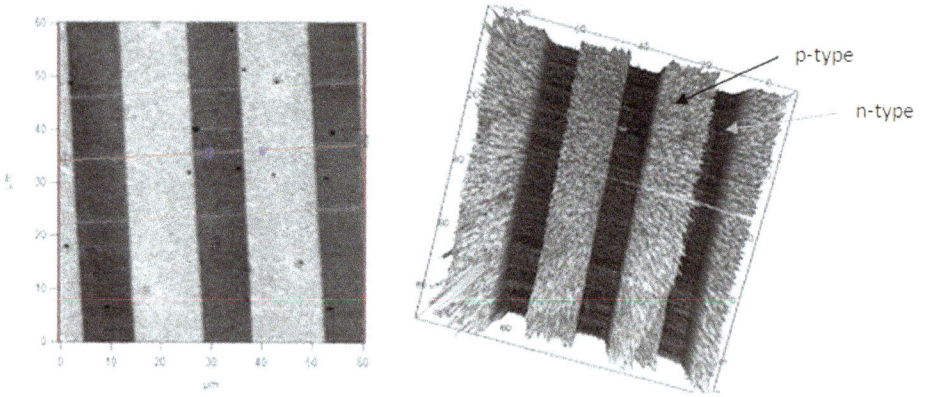

Fig. 4. 2-D and 3-D Profile of stack.

This evidence supports that the QDSL formation using Silicon and Germanium QDs has a rather uniform distribution and appropriate height. Since the self-assembly process is essential for device integration, confirming technique and procedure of self assembly is important to producing accurate results.

It is worth noting that the line sample has silicon dots assembled first and then germanium dots. It is most optimal to assemble higher band gap material on top of lower band gap material for the best multi state mini band transition. Assembling Si dots and then Ge dots is not the most optimal. Creating this structure was easier to assemble in four layers because the germanium dots tend to assemble easier when more cladding is present. In addition, annealing silicon and then germanium dots follows a temperature profile of 750° and then 425°. Eventually collecting AFM data from assembling silicon dots on germanium dots will yield more separated intermediate states.

The AFM data is very important in proving that the QDSL self-assembly, QDs size, and QDSL distribution is uniform and repeatable. Having confirmed the dot size, we can use the same methods to deposit the dots in this formation onto a device.

3. Possible Applications in QDG FETs

The use of dual layered dots can be applied to a quantum dot gate field effect transistor (QDG-FET). With these quantum dots, there exist multiple intermediate drain current thresholds in between the "ON" and "OFF" state [6]. This is known as multi-value logic (MVL) [7, 8]. The use of QDSL can help make two-bit MOSFETs. These MOSFETs exhibit multiple states through tunneling from channel to retaining charge within the quantum dot layers. Using multiple self-assemblies of dots of varying size allows for different charge thresholds represented in the transitioning in states.

A QDG FET is a conventional FET with quantum dots being deposited along the gate region. Figure 5a and b represent a QDG with one self-assembly. The quantum dots within this device confine electrons allowing for discrete energy levels to create wells and barrier between the outer cladded layers and the core [4]. The charge retention within the inner core is important to multi-bit processing.

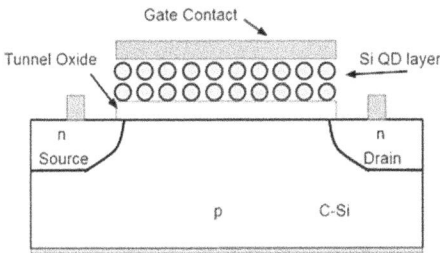

Fig. 5a. Si QDG Cross section.   Fig. 5b. Ge QDG Cross section.

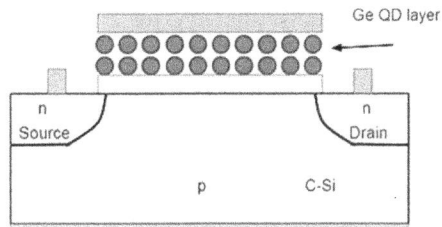

Silicon and Germanium quantum dots have been shown to retain charges allowing for transistors to show multi-step bit functionality. This means device sizes can maintain smaller sizes while increasing memory density. The charges tunnel from the channel region into the quantum dots where the charges are stored within this quantum dot gate region. During the intermediate states, the drain current remains stagnant until the QD mini-bands are fully charged. This is due mainly in part to the mini-bands created by the quantum dots. Mini-band structures are energy bands that overlap one another in the same structure as quantum dots [6]. These mini-bands are able to retain charges within the cores of the quantum dots. Then the charges start to tunnel into the gate. Since the quantum dots have a cladded layer, the likely-hood of charge retention is much higher [2, 9]. Mini-band properties are important towards creating multi-state FET devices [10] where electrons move through the QD layers creating intermediate states.

Figure 6 shows the theoretical I_D-V_G for a 4-state QDG-FET device. The intermediate states (i_1 and i_2) occur when the drain current stagnates. When the gate voltage is large enough, the charge carriers will tunnel from the cladded quantum dot layers into the min-energy bands between the conventional ON and OFF states [2]. Charges continue to fill the mini-bands until it is completely occupied. When this occurs, the charges will move and transfer to the next highest mini-band leading to an incremental increase in I_D until

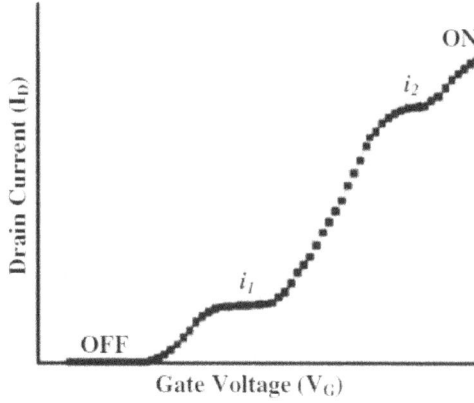

Fig. 6. Theoretical Id-Vg Curve for a four state QDG [2].

the entire bottom QD layer is filled [2]. This will cause charges to transfer to upper layers of quantum dots, allowing for a shift in states from i_1 to i_2. The drain current can be represented by:

$$I_D = \frac{W}{L} * c_{ox} * \mu_n \left[(V_G - V_{TH})V_{DS} - \frac{V_{DS}^2}{2} \right] \tag{1}$$

Where W and L represent width/length dimensions of the channel, μ_n is the electron mobility, and c_{ox} represents the oxide capacitance. The shift in different states can also be observed using mini-band transitions.

Figure 7 shows the carrier vs. Fermi level simulated plot. This is used to quantify the amount of charge that can be stored within the conduction mini-band for QDSLs. This plot is important for showing the mini-band transitions. Each transition from a stable carrier density represents the charges moving to the next mini-band. Silicon QDs have shown to have more transitions and more mini-bands compared to germanium QDSLs because silicon is larger in size and thus has a larger band gap compared to germanium. Since germanium is also smaller in size, the charges reach a saturated state much earlier than silicon. This would imply that silicon dots would be able to store more charges compared to germanium.

It is important to note that size of dots or cladding layer affects the separation changing the overall voltage required to arrive at the intermediate states [2]. Using two different materials for the dots allows for easy creation of dots of different sizes. These intermediate states allow multi-value logic for overall reduced transistor count and higher memory density. Using a QDSL structure allows for multiple varying mini-energy bands that enable control of threshold voltages. QDG-FETs can be created using CMOS fabrication techniques, leading to flexible design.

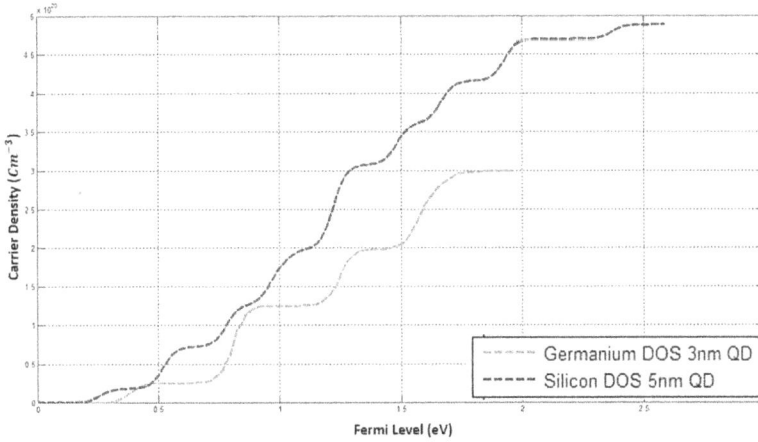

Fig. 7. QD Mini-band Transition.

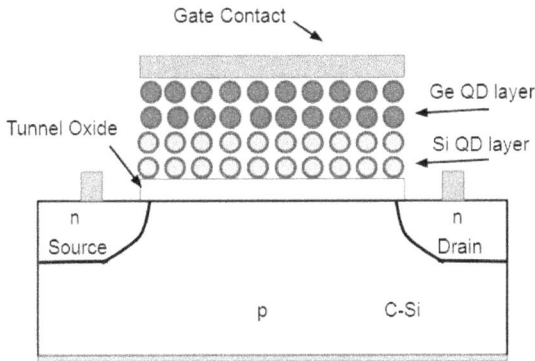

Fig. 8. QDSL QDG cross section.

Using the AFM imaging and self-assembly techniques from earlier, a QDG with mixed dot QDSL could be created in the configuration of Fig. 8. The goal with using two different sized dots in the QDSL structure is to create wider intermediate states compared to previous iterations of a mixed dot QDSL QDG-FET. By separating the layers of different sized dots, larger mini-bands need to be charged resulting in larger regions of charge when moving from one state to another state. Mixed dots in the structure of Fig. 8 allows for more distinct drain current contrast.

4. Discussion

When the gate voltage is increased, it increases the drain current. An equation for drain current in a conventional NMOS device was represented in equation 1.

When working with QDG based devices, the quantum dots change the characteristics of the drain current equations because the charges can be stored within QDs and tunnel between QDs. Equation 2 is a model representing how the size, quantity and number of layers of quantum dots affect the overall threshold voltage [11].

$$
V_{Th} = -\frac{q}{C_0''}\left[\left[\sum_{i=1}^{2}\frac{x_{QD(Ge)}n_{Ge}N_{QD(Ge)}}{x_g}\right] + \left[\sum_{i=3}^{4}\frac{x_{QD(Si)}n_{Si}N_{QD(Si)}}{x_g}\right]\right] \tag{2}
$$

Where $x_{QD(Ge)}$ and $x_{QD(Si)}$ are the distance of quantum dot core for the germanium and silicon dots. The number of dots is represented as n_1 and n_2 in the multiple layers. $N_{QD(Ge)}$ and $N_{QD(Si)}$ are the charges on the silicon and germanium dots. C_0'' represents the oxide capacitance and x_g is the distance of the insulating layer from the gate, with q being the electron charge. As the charge within the quantum dots increases, it increases the overall V_{Th}. This equation can be related to a quantum dot model for drain current in equation 3 [11].

$$
I_d = \frac{W}{L} * C_0''\mu_n\left[\{V_g - (V_{TH1} + \Delta V_{TH1})\}V_d - \frac{V_d^2}{2}\right] \tag{3}
$$

Where V_{TH1} represents the first well's voltage threshold and μ_n is the electron mobility. It can be seen that as the number of varying dots in size and quantity increases, the threshold voltage increases which has a positive correlation with drain current. The intermediate states occur when the charge carriers undergo tunneling between the multiple QD layers [2-4, 7, 9, 11]. The increase of threshold voltage occurs when the charge transfers from the band gap well to the quantum dot layer. This shift can be even more pronounced when different sizes of dots are used because then more voltage is required to move the charges out of the larger wells.

Previous devices have shown the capability for mixed dot QDG devices with multiple intermediate states [2, 9]. Figures 9 and 10 shows the cross section device and the output I_d-V_g characteristics of a previously successfully fabricated mixed dot QDSL FET.

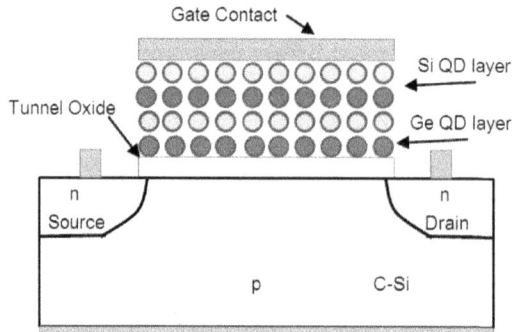

Fig. 9. Mixed 4 layer QDSL.

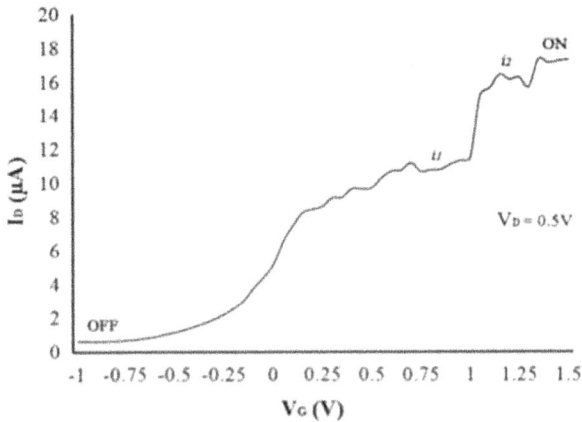

Fig. 10. I_D vs. V_G for Mixed QDSL [2].

Using QDSL structures creates discrete mini-bands. These mini-bands require varying amounts of threshold voltage and vary in size with the size of the quantum dots. These varying thresholds lead to the creation of intermediate states. Each time the drain current stabilizes; the charges are stored within the quantum dots. These charges begin to leak when they have fully charged the inner core [5]. Each time two layers of dots have been charged, it is represented by a new intermediate state. By using two different sized dots, four state devices can be achieved. This will help with the advancement of QDG-FET devices mentioned previously as well as NVM devices. QDG-NVM devices can mimic flash memory floating gate capabilities.

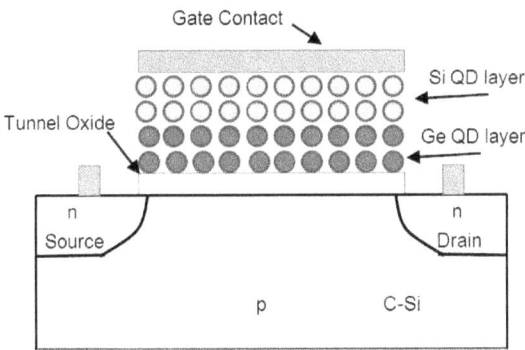

Fig. 11. Ideal mixed dot QDSL Structure.

The ideal structure for a QDSL FET is represented by Fig. 11. This figure shows germanium to be assembled on the lower level with silicon dots assembled on top. This is more effective because the quantum dots will require increasingly larger amounts of voltage to tunnel past smaller and then larger dots. This will result in more distinct separated states with wider separation compared to previous models. This surface profile has yet to be demonstrated using AFM, and will be more difficult due to the material layering. Annealing temperatures of germanium tend to be rather low, while silicon annealing temperatures are much higher. Applying a higher temperature during the second self assembly of silicon could have an effect on the germanium so a lower secondary annealing temperature of 425°C is taken when assembling the second layer.

Using these designs as a template, FET devices can easily be created with the goal of trying to achieve more separated device states. Trapping these charges within distinct mini-bands leads to easier distinction of states leading to more stability and device endurance.

5. Conclusion

Moving forward, multi-dot QDSL layers show lots of promise for NVM and FET devices. This is mainly due to their ability to store and retain charges and to be dispersed uniformly over select regions. Easy tunability can also lead to some varying electrical properties. It would also be interesting to characterize QDSL device characteristics using different sized dots. This would entail, using two self assemblies of varied sized silicon dots. This can be done for varying the etch times during the creation of the self assembly solution.

QDSL structures should also be noted to have the potential for photonic applications in solar cells. Multi-junction solar cells have been growing as a more efficient solar cell. If dual layers of different dots could be deposited onto solar cells, then it has the potential to create a solar cell with an absorption spectrum. QDSL based solar cells can imitate the varying band gap and absorption spectrums that are also exhibited in Multi-junction solar cells [12].

The design flexibility of QDSLs makes them very useful material for semiconductor device design. Moving forward, the electrical characteristics of dual dot QDSL FETs will be collected and are expected to exhibit wider intermediate state separation.

References

1. T. Phely-Bobin, D. Chattopadhyay, and F. Papadimitrakopoulos, "Characterization of mechanically attrited Si/SiOx nanoparticles and their self-assembled composite films," *Chem. Mater.* 14(3), pp. 1030-1036, March 2002.
2. M. Lingalugari *et al.*, "Novel Multistate Quantum Dot Gate FETs Using SiO2 and Lattice-Matched ZnS-ZnMgS-ZnS as Gate Insulators," Journal of ELECTRONIC MATERIALS, Vol. 42, No. 11, 2013.
3. F. C. Jain, E. Suarez, M. Gogna, F. Alamoody, D. Butkiewicus, R. Hohner, T. Liaskas, S. Karmakar, P.-Y. Chan, B. Miller, J. Chandy, and E. Heller, "Novel Quantum Dot Gate FETs and Nonvolatile Memories Using Lattice-Matched II–VI Gate Insulators," *Journal of Electronic Materials*, Vol. 38, Issue 8, pp. 1574-1578, Aug. 2009.
4. M. Gogna, E. Suarez, P.-Y. Chan, F. Al-Amoody, S. Karmakar, and F. Jain, "Nonvolatile silicon memory using GeOx-cladded Ge quantum dots self-assembled on SiO2 and lattice-matched II–VI tunnel insulator," *Journal of Electronic Materials*, Vol. 40, Issue 8, pp. 1769-1774, Aug. 2011.

5. M. Lingalugari, P.-Y. Chan, E. K. Heller, and F. C. Jain, "Multi-bit quantum dot nonvolatile memory (QDNVM) using cladded germanium and silicon quantum dots," *International Journal of High Speed Electronics and Systems*, Vol. 24, No. 03n04, pp. (1550003-1)-(1550003-11), Sep. & Dec. 2015.

6. S. Karmakar, F. C. Jain, "Future Semiconductor Devices for Multi-Valued Logic Circuit Design" Materials Sciences and Applications," August 2012, pp. 807-814.

7. L. J. Michael, A. H. Taddiken and A. C. Seabaugh, "Multiple-Valued Logic Computation Circuits Using Micro and Nanoelectronics Devices," Proceedings of 23rd IEEE International Symposium on Multiple-Valued Logic, Sacramento, 24-27 May 1993, pp. 164-169.

8. S. Karmakar *et al.*, "Fabrication and Circuit Modeling of NMOS Inverter Based on Quantum Dot Gate Field-Effect Transistors," Journal of Electronic Materials, Vol. 41, No. 8, August 2012.

9. S. Karmakar, E. Suarez, F. C. Jain, "Three-State Quantum Dot Gate FETs Using ZnS-ZnMgS Lattice-Matched Gate Insulator on Silicon," Journal of ELECTRONIC MATERIALS Vol. 40, No. 8, 2011.

10. F. Jain, S. Karmakar, P.-Y. Chan, E. Suarez, M. Gogna, J. Chandy, E. Heller, *Journal of Electronic Materials*, 2012, Vol. 41(10), pp. 2775-2784.

11. F. C. Jain, B. Miller, P.-Y. Chan, S. Karmakar, F. Al-Amoody, M. Gogna, J. Chandy, E. Heller, "Spatial Wavefunction-Switched (SWS) InGaAs FETs with II-VI Gate Insulators," *Journal of Electronic Materials*. 40(8), pp. 1717-1725, May 2011.

12. X. Lan, O. Voznyy, F. P. Garcia de Arquer, M. Liu, J. Xu, A. H. Proppe, G. Walters, F. Fan, H. Tan, M. Liu, Z. Yang, S. Hoogland, "10.6% Certified Colloidal Quantum Dot Solar Cells via Solvent-Polarity-Engineered Halide Passivation," *Nano Letters*, 16, pp. 4630-4634, June 2016.

Circuits and Simulation of Quaternary SRAM Using Quantum Dot Channel Field Effect Transistors (QDC-FETs)

Bander Saman

Department of Electrical Engineering, Taif University,
P.O. BOX 888 – 21974 – Hawiyah - Taif - KSA
Bander.Saman@uconn.edu

Jun Kondo[*,1], J. Chandy[†,1], and F. C. Jain[‡]

Department of Electrical and Computer Engineering, University of Connecticut,
371, Fairfield Way, U-2157, Storrs, CT, 06269-2157, USA
[*]kondo@uconn.edu
[†]john.chandy@uconn.edu
[‡]faquir.jain@uconn.edu

This paper presents the design and simulation of static random access memory (SRAM) using Quantum Dot Channel Field-Effect Transistors (QDC-FETs). A QDC-FET consist of two quantum dots (3 nm to 4 nm) forming n-channel between the source and drain Quantum Dot Channel (QDC) on a p-Si substrate regions. The quantum dot channel enables higher-mobility transport on very low-mobility substrates. The structure of a quantum dot channel QDC-FET that has shown four-state characteristics of charge carriers from one channel to other channel of the device. It shows that four states can be obtained in an inverter made of four-state QDC-FETs. The four stats QDC-FETs is suitable for quaternary logic application with less complex and more area efficient than existing quaternary logic circuits. The device with four-state has been modeled using Berkeley Short-channel IGFET Model (BSIM) and Analog Behavioral Model (ABM), the model is suitable for transient analysis at circuit level. This model is optimized for a quaternary inverter logic and used to replace a conventional CMOS SRAM.

Keywords: QDCFETs; multi-state FETs; 2 bit SRAM; VLSI.

1. Introduction

Figure 1 shows a cross-sectional schematic of a fabricated SiOx-cladded Si quantum dot channel (QDC) Si FET having two quantum dots in the gate region (QDG). I_{DS}–V_{GS} and I_{DS}–V_{DS} characteristics of a SiOx-Si QDC-FET are shown in Fig. 2. The characteristics show intermediate states (three-state) as a function of the drain voltage [1, 2].

---

[1]Corresponding authors.

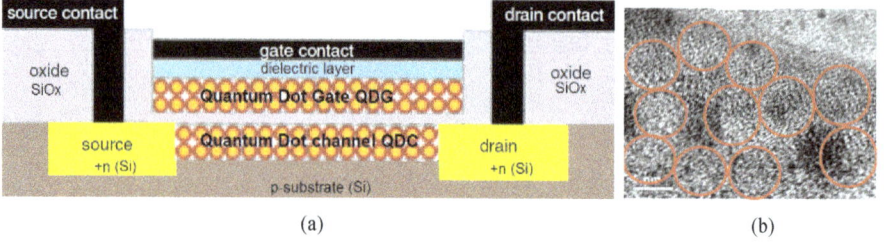

(a)   (b)

Fig. 1. (a) SiOx-Si QDC- FET having QDG. (b) TEM showing four layers of SiOx-Si QDG and QDC [2].

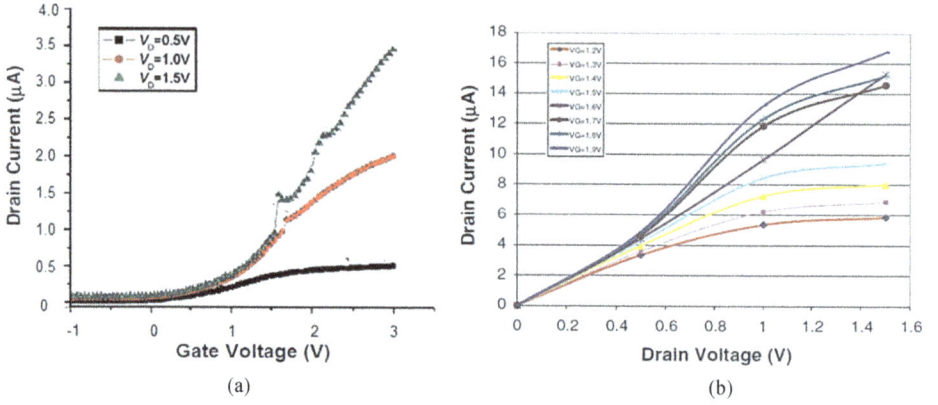

(a)   (b)

Fig. 2. (a) I_D-V_G and (b) I_D-V_D characteristics of a SiOx-Si QDC-FET having QDG [1].

2. QDC-FET Circuit Model

A QDC-FET behaves as a MOS-FET. Eq. (1) illustrates the drain current for the MOS-FET, the same equation was developed in Eq. (2) for the QDC-FET to represent the drain current with intermediate states [1]. A changing in the gate voltage (VG) responds to a shift in the threshold voltage (ΔV_{TH}) as showing in Eq. (3) [1-3].

For three-state device, ΔV_{TH} is divided into three regions corresponding to the three regions as described in Eq. (4), the region 1 or OFF state (when VG<Vg1), region 2 or intermediate state i (when Vg1<VG<Vg2) and region 3 or ON state (when VG>Vg2) [1, 3].

In four states n-QDC-FET, the threshold voltage changes linearly when the gate voltage increases through a range of voltages (Vg1 to Vg2, Vg3 to Vg4), this showing in Eq. (5) [1, 2]. According to this equation system, the parameters of Vg1, Vg2, Vg3, and Vg4 thresholds voltage can be adjusted by the size of the dots.

$$I_{DS} = \left(\frac{W}{L}\right) C_{OX}\mu_n \left((V_{GS} - V_{TH})V_{DS} - \frac{V_{DS}^2}{2} \right) \tag{1}$$

$$I_{DS} = \left(\frac{W}{L}\right) C_{OX}\mu_n \left((V_{GS} - V_{TH-EFF})V_{DS} - \frac{V_{DS}^2}{2}\right) \tag{2}$$

$$V_{TH-EFF} = V_{TH} + \Delta V_{TH} \tag{3}$$

$$\Delta V_{TH} = \begin{cases} 0, V_{GS} \le Vg1, \textbf{State OFF} \\ \alpha 1(V_{GS} + Vg1), Vg1 < V_{GS} \le Vg2, \textbf{State i} \\ 1(Vg2 + Vg1), V_{GS} > Vg2, \textbf{State ON} \end{cases} \tag{4}$$

$$\Delta V_{TH} = \begin{cases} 0, V_{GS} \le Vg1, \textbf{OFF} \\ \alpha 1(V_{GS} + Vg1), Vg1 < V_{GS} \le Vg2, \textbf{MOD1} \\ \alpha 1(Vg2 + Vg1), Vg2 < V_{GS} \le Vg3 \\ \alpha 1(Vg2 + Vg1) + \alpha 2(V_{GS} + Vg3), Vg3 < V_{GS} \le Vg4, \textbf{MOD2} \\ \alpha 1(Vg2 + Vg1) + \alpha 2(Vg4 + Vg3), V_{GS} \ge Vg4, \textbf{ON} \end{cases} \tag{5}$$

Where

W, L   are the channel width and the channel length

C_{OX}, μ_n are the gate capacitance per unit area and the channel mobility

V_{GS}, V_{DS} are the gate to source voltage and the drain to source voltage

V_{TH} is the threshold voltage

V_{TH-EFF} is the effective threshold voltage

ΔV_{TH} is the threshold voltage shift

Vg1   is threshold voltage of state 1

Vg2   is threshold voltage of state 2

Vg3   is threshold voltage of state 3

Vg4   is threshold voltage of state 4

Berkeley Short-channel IGFET Model (BSIM3) and Analog Behavioral Model (ABM) libraries are used to establish n-type QDC-FET model that can address the changing in threshold voltage ΔVT [4, 5]. The output of ABM block represents V_G-ΔVT, the applied gate voltage of the BSIM transistor is equal to the output of ABM block, and the threshold voltage of the BSIM transistor is VT. This approach captures the behavior of the four states n-QDC-FET. The advantage of using BSIM models is that they can scale to 45 nm and 25 nm with capturing the latest technology advances [6].

This model is set in hierarchical block and it ready to be used in Cadence-OrCAD CIS as shown in Fig. 3. The I_{DS}-V_{GS} characteristic of the four states n-QDC-FET is shown in Fig. 4, the device parameters are (L=1um, W=5um, V_{DD}=3V, VT=0.5, Vg1=1/3*V_{DD}=1V, Vg2=1.5/3*V_{DD}=1.5V, Vg3=2/3*V_{DD}=2V, and Vg4=2.5/3*V_{DD}=2.5V). The simulation result shows the four regions of the transfer characteristics (OFF-MOD1-MOD2-ON).

QDC

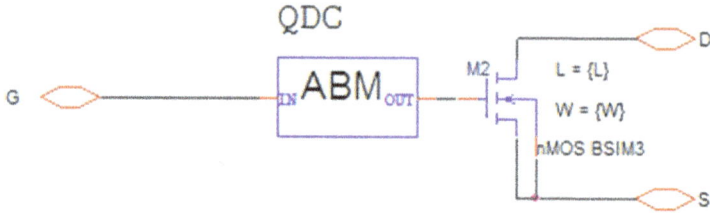

Fig. 3. For states n-QDC-FET model circuit.

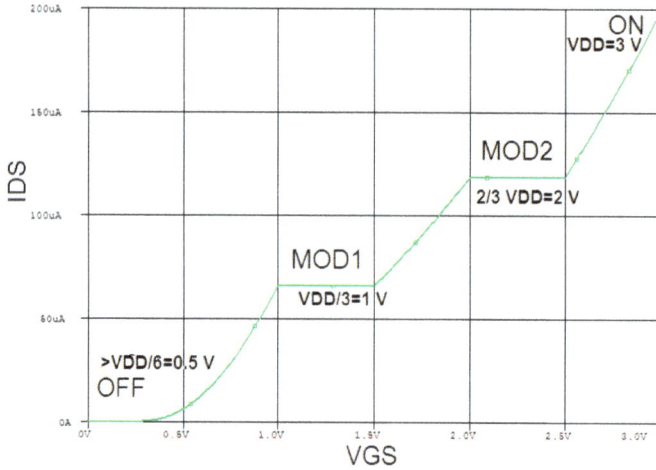

Fig. 4. 1um four states n-QDC-FET I_{DS}-V_{GS} characteristic.

3. QDC-FET Inverter Logic Gates.

The four stats QDC-FETs is suitable for quaternary logic application with less complex and more area efficient than existing quaternary logic circuits [7, 8].

The quaternary inverter logic circuit and transfer characteristics are shown in Fig. 5 and Fig. 6, respectively. Figure 7 shows the transient time simulation of the quaternary inverter logic, where the quaternary input signal has four voltage level (0=Logic0, $\frac{1}{3}V_{DD}$=Logic1, $\frac{2}{3}V_{DD}$=Logic2, V_{DD}=Logic3) [1-3].

The four states inverter comprising of p-MOS and four states n-QDC-FET (pseudo-nMOS). If $V_{IN} < \frac{1}{3}V_{DD}$ (Logic0), the output of the inverter is $\approx V_{DD}$ (Logic3). When V_{IN} in between $\frac{1}{3}V_{DD}$ and $\frac{2}{3}V_{DD}$ (Logic1), the output of the inverter is $\frac{2}{3}V_{DD}$ (Logic2). The inverter output is $\frac{1}{3}V_{DD}$ (Logic1) when $V_{IN} \approx \frac{2}{3}V_{DD}$ (Logic2). The output of the inverter is $< \frac{1}{3}V_{DD}$ (Logic0) if $V_{IN} > \frac{2}{3}V_{DD}$ (Logic3). The truth table of the quaternary inverter logic is shown in Table 1 [1-3].

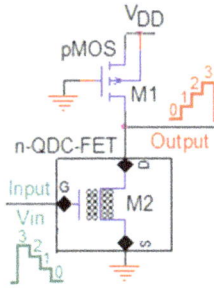

Fig. 5. The four states n-QDC-FET inverter circuit.

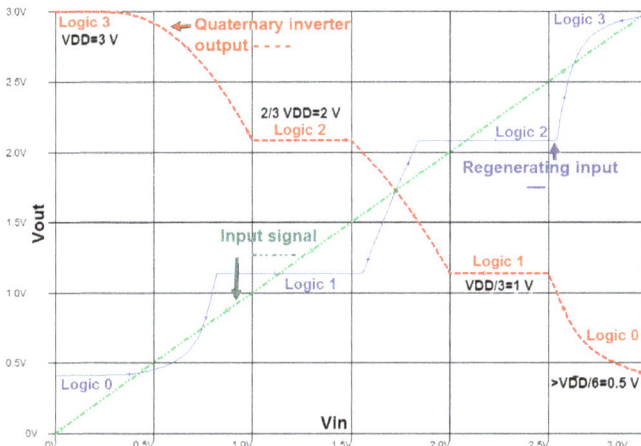

Fig. 6. The transfer characteristic of the four states n-QDC-FET inverter.

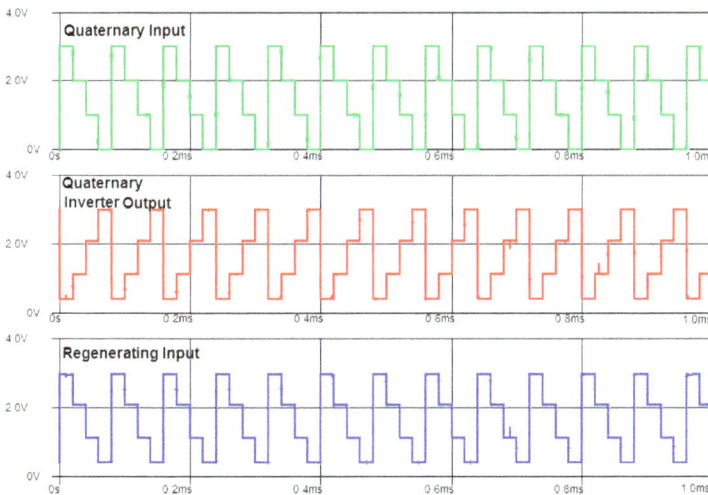

Fig. 7. The F The transient time simulation of the quaternary inverter logic.

The input single is regenerated simply by inverting the output of the circuit in Fig. 5 using additional four states inverter, the regenerating input signal is showing in Fig. 6 [6].

Table 1. The truth table of the quaternary inverter logic.

Quaternary Inverter Input		Quaternary Inverter Output	
Voltage Level	Logic	Voltage Level	Logic
0	0	$\approx V_{DD}$	3
$\frac{1}{3}V_{DD}$	1	$\approx \frac{2}{3}V_{DD}$	2
$\frac{2}{3}V_{DD}$	2	$\approx \frac{1}{3}V_{DD}$	1
V_{DD}	3	$< \frac{1}{6}V_{DD}$	0

4. SRAM Circuit Model

A Conventional 1 bit SRAM consists of two conventional-coupled COMS Inverter for the memory implementation. The 6T circuit has two coupled inverter and two n-MOSFET access transistors as shown in Fig. 8(a) [9]. M1, M2, M3 and M4 transistors make a pair of inverters, connected in a loop. The other transistors M5 and M6 are used to control read and write. The 8T circuit of Fig. 8(b) has two more n-MOS to prevent the data from being disturbed during a read operation, the truth table of 1 bit SRAM shown in Table 2 [9-10].

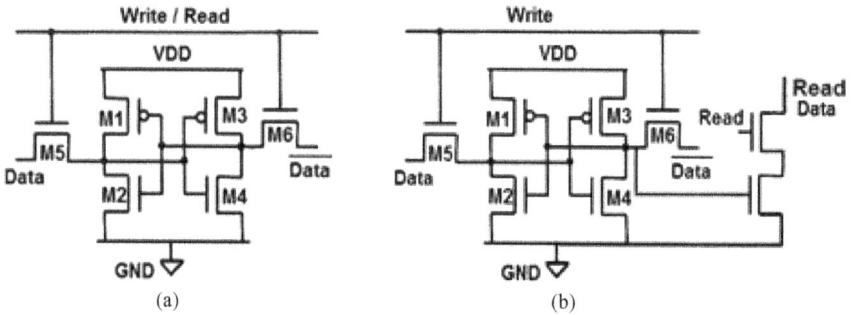

Fig. 8. CMOS 1 bit SRAM (a) 6T cell and (b) 8T cell.

Table 2. CMOS 1 bit SRAM truth table.

Write signal	Data signal	Stored Data
0	0	Same State
0	1	Same State
1	0	0
1	1	1

The same circuit can be used to make 2 bit SRAM by replacing the conventional coupled COMS inverter with two n-QDC-FET inverter as showing in Fig. 9. Where the data line is a quaternary input signal has four voltage level (0=Logic0, $\frac{1}{3}V_{DD}$=Logic1, $\frac{2}{3}V_{DD}$=Logic2, V_{DD}=Logic3) and the word line (WR) is a binary signal that enabling and disabling the write operation. The quaternary data is stored by two coupled inviters and while the write signal is 0. When the write signal is 1, the coupled inviters change their value according the data single.

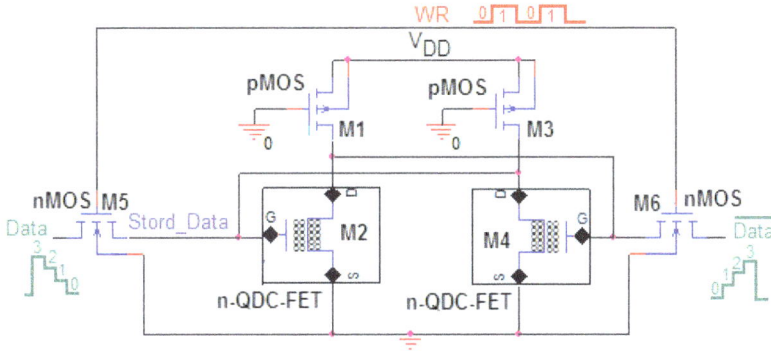

Fig. 9. The two bit SRAM 6T cell using n-QDC-FET inverter.

Figure 10 shows transient simulations of 2 bit SRAM 6T cell and the parameters of the cell are shown in Table 3. In case of Write input signal is on (=VDD), Stored Data signal flows Data input signal. When Write input signal is switched off (=0), the voltage/logic stays at the storage node. The simulation result depicts the 2 bit SRAM stored the data as expected and the truth table of 2 bit SRAM (Table 4).

Table 3. 2 bit SRAM 6T cell parameters.

Parameter	Value
p-MOS M1 and M3 Length L	1 μm
p-MOS M1 and M3 Width W	10 μm
n-QDC-FET M2 and M4 Length L	1.0 μm
n-QDC- FET M2 and M4 Width W	5 μm
V_{DD}	3 V
p-MOS V_{TH}	-0.6 V
n-MOS V_{TH} and n-QDC- FET V_{TH}	0.5 V
n-QDC- FET Vg1	1.0/3 * VDD=1.0 V
n-QDC- FET Vg2	1.5/3 * VDD=1.5 V
n-QDC- FET Vg3	2.0/3 * VDD=2.0 V
n-QDC- FET Vg4	2.5/3 * VDD=2.5 V

Table 4. The truth table of 2 bit SRAM.

Write signal	Data signal	Stored Data
0	0 V = Logic 0 (00)	Same State
0	⅓ V_{DD} = Logic 1 (01)	Same State
0	⅔ V_{DD} = Logic 2 (10)	Same State
0	V_{DD} = Logic 3 (11)	Same State
1	Logic 0 (00)	> ⅙ V_{DD} = Logic 0 (00)
1	Logic 1 (01)	⅓ V_{DD} = Logic 1 (01)
1	Logic 2 (10)	⅔ V_{DD} = Logic 2 (10)
1	Logic 3 (11)	V_{DD} = Logic 3 (11)

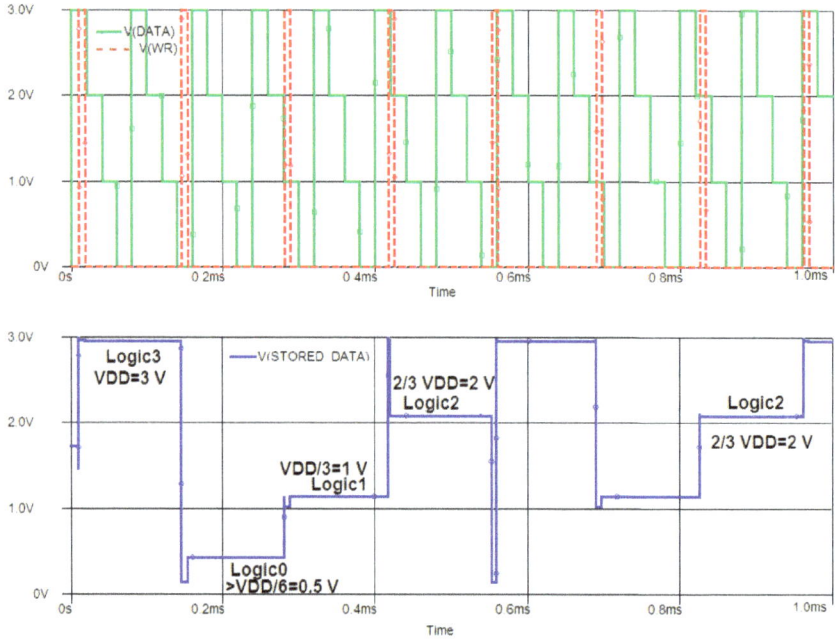

Fig. 10. The simulation of 2-bit SRAM

5. Conclusions

In this paper, the simulations of I_{DS}-V_{GS} characteristics present of four state quantum dots channel (QDC) n-channel FET. Also, QDC-FET is used to design the quaternary inverter logic and 2-bit SRAM cell. The transient simulations of the quaternary signal present to verify the functionality of the circuits. The two cascading inviters regenerate the quaternary input signal and the coupled inviters allow the signal to be stored. The 2-bit SRAM cell based on QDC-FETs offers the write and the read operation as the CMOS 1 bit SRAM 6T.

References

1. Jain, Karmakar, Chan, Suarez, Gogna, Chandy, and Heller. (2012). Quantum Dot Channel (QDC) Field-Effect Transistors (FETs) Using II–VI Barrier Layers. Journal of Electronic Materials, 41(10), 2775-2784.
2. F. Jain, P.-Y. Chan, E. Suarez, M. Lingalugari, J. Kondo, P. Gogna, B. Miller, J. Chandy, and E. Heller. (2013). Four-State Sub-12-nm FETs Employing Lattice-Matched II–VI Barrier Layers. Journal of Electronic Materials, 42(11), 3191-3202.
3. Karmakar, S. (2013). Design of four-state inverter using quantum dot gate-quantum dot channel field effect transistor. Electronics Letters, 49(18), 1131-1133.
4. BSIM3v3 Manual, (Final Version), web site: rely.eecs.berkeley.edu or 128.32.156.10.
5. Analog Behavioral Modeling Applications reference manual. Cadence – application note, December 2009.
6. B. Saman, P. Mirdha, M. Lingalugari, P. Gogna, and F. C. Jain. Logic Gates Design and Simulation Using Spatial Wavefunction Switched (SWS) FETs. International Journal for High-Speed Electronics and Systems, Vol. 24, Nos. 3 & 4 (2015) 1550008
7. Gogna, Suarez, Lingalugari, Chandy, Heller, Hasaneen, and Jain. (2013). Ge-ZnSSe Spatial Wavefunction Switched (SWS) FETs to Implement Multibit SRAMs and Novel Quaternary Logic. Journal of Electronic Materials, 42(11), 3337-3343.
8. Temel, T. and Morgul, A. (2002). Multi-valued logic function implementation with novel current-mode logic gates. Circuits and Systems, 2002. ISCAS 2002. IEEE International Symposium on, Vol. 1, pp. I-I.
9. James Boley, Jiajing Wang and Benton H. Calhoun, Analyzing Sub-Threshold Bitcell Topologies and the Effects of Assist Methods on SRAM Vmin, JLPEA 2012, 143-154.
10. Evelyn Grossar, Michele Stucchi, Karen Maex, and Wim Dehaene, Read stability and write-ability analysis of SRAM Cells for nanometer Technologies, IEEE Journal of Solid-State Circuits, Vol. 41, No. 1, 2006.

Modeling and Fabrication of GeO$_x$-Ge Cladded Quantum Dot Channel (QDC) FETs on Poly-Silicon

Jun Kondo*, Pial Mirdha, Barath Parthasarathy, Pik-Yiu Chan, Bander Saman, and Faquir Jain†

Electrical and Computer Engineering, University of Connecticut,
371 Fairfield Way, Unit 4157, Storrs, CT 06269, USA
*jun.kondo@uconn.edu
†faquir.jain@uconn.edu

Evan Heller

Synopsis Inc., Ossining, NY 10562, USA
evankheller@gmail.com

Quantum dot channel (QDC) and Quantum dot gate (QDG) field effect transistors (FETs) have been fabricated on crystalline Si using cladded Si and Ge quantum dots. This paper presents fabrication and modeling of quantum dot channel field effect transistors (QDC-FETs) using cladded Ge quantum dots on poly-Si thin films grown on silicon-on-insulator (SOI) substrates. HfAlO$_2$ high-k dielectric layers are used for the gate dielectric. QDC-FETs exhibit multi-state I-V characteristics which enable two-bit processing, and reduce FET count and power dissipation. QDC-FETs using germanium quantum dots provide higher electron mobility than conventional poly-silicon FETs, and mobility values comparable to conventional FETs using single crystalline silicon.

Keywords: Quantum dot channel; QDC-FET; cladded Ge quantum dots; HfAlO$_2$ high-k dielectric.

1. Introduction

This paper presents experimental I_D-V_G and I_D-V_D characteristics of GeO$_x$-Ge cladded quantum dot channel (QDC) FETs on poly-silicon. Their operation is modeled by carrier transport in narrow energy mini-bands which are manifested in a quantum dot superlattice (QDSL) transport channel [1]. QDSL is formed by an array of cladded Germanium quantum dots (GeO$_x$-Ge). Multi-state FETs are needed in multi-valued logic (MVL) that can reduce the number of gates and transistors in digital circuits [2].

2. Structure

The structure of a QDC FET is shown in Figs. 1(a) and 1(b). The QDC FET consists of two or more layers of cladded germanium quantum dots in the channel regions. Here, GeO$_x$ cladding forms the thin-barrier ($\sim$1nm) over Ge quantum dots (3nm in diameter). An array of these cladded dots behaves as a quantum dot superlattice (QDSL). This superlattice has

---

†Corresponding author.

energy mini-bands that are very narrow and separated with larger energy than in conventional quantum well/wire superlattices. The Kronig-Penney model was used to determine the energy mini-band locations and widths [1]. HfAlO$_2$ nanolaminate high-k dielectric combinational layers were used for the tunnel oxide of the transistor in order to attain a relatively large dielectric constant. The combinational layers were deposited using the Savannah Atomic Layer Deposition (ALD) system at Harvard University. The measured QDC FET device has the gate length and the gate width of 60μm, and a gate width to gate length ratio (W/L ratio) of 26/25.

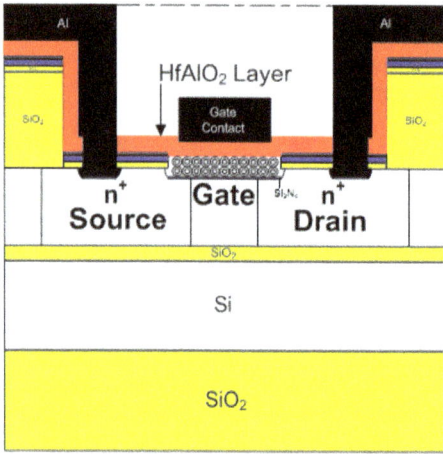

Fig. 1(a). Structure of QDC FET.   Fig. 1(b). Gate Structure of QDC FET.

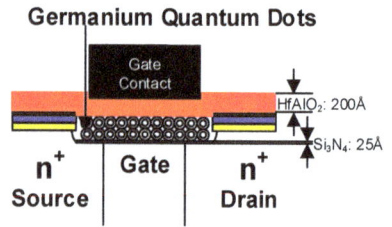

3. Characteristics

The fabricated QDC FET was tested for multi-state characteristics. Figure 2 shows the I_D-V_G characteristics of QDC FET. As V_G was changed from -3 to 13 Volts, V_D changed from 0 to 14 Volts. When V_D was greater than 2 Volts, I_D peaks was observed at numerous locations. The distinct three-state characteristic (low-intermediate-high) were observed when V_D was equal to 12, 14 Volts.

Figure 3 shows the I_D-V_D characteristics. These I_D-V_D characteristics represent three distinct groupings. V_G less than 8.5 Volts formed the first group, V_G equal to 8.5, 9, 9.5, 10, 10.5 Volts formed the second group, and the V_G equal to 11, 11.5, 12, 12.5, 13 Volts formed the third group. The distinct groupings were also observed in the I_D-V_D characteristics of quantum dot gate–quantum dot channel (QDG-QDC)FETs. I_D-V_D characteristics of QDC FETs were completely different from QDG-QDC FETs.

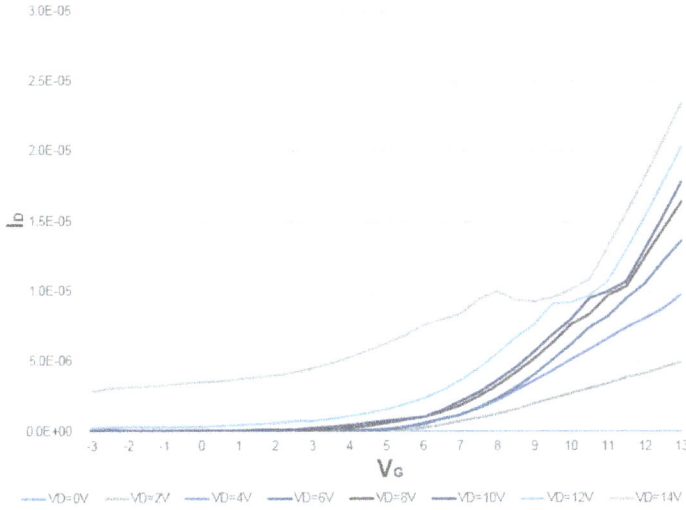

Fig. 2. I$_D$-V$_G$ characteristics of GeO$_x$-Ge quantum dot channel (QDC) FET on SOI substrate.

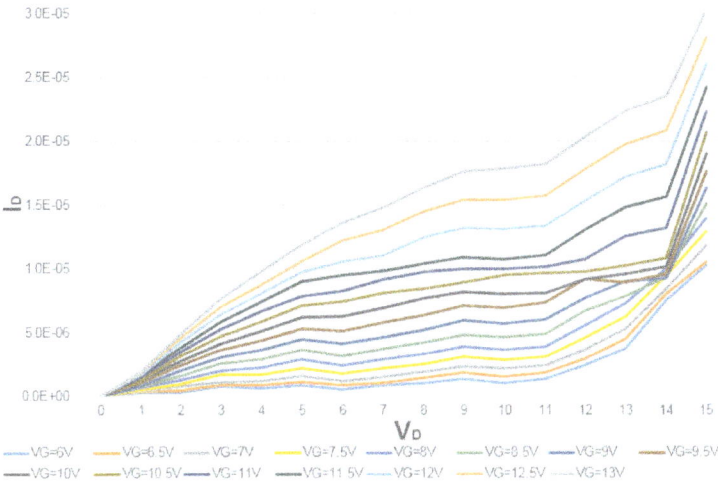

Fig. 3. I$_D$-V$_D$ characteristics of GeO$_x$-Ge quantum dot channel (QDC) FET on SOI substrate.

4. Quantum Simulations

The Kronig-Penney model [1] was used to determine the energy mini-band locations and their widths in the quantum dot superlattice (QDSL) formed in the array of thin-barrier (~1nm) cladded Ge quantum dots, and these mini-energy bands are shown in Fig. 4. The electron transport in the inversion channel is influenced by the energy mini-bands manifested due to the quantum dot channel.

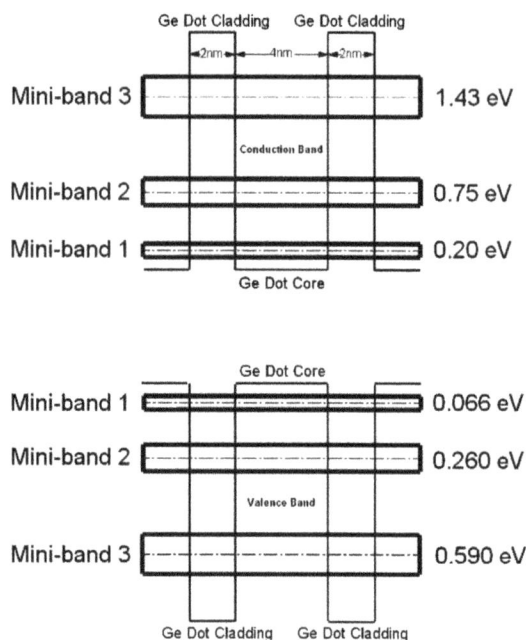

Fig. 4. Energy mini-band Locations in the Ge dot QDSL.

Figures 5(a) and 5(b) show the simulation of electron wavefunction in GeO_x-Ge quantum dot superlattice (QDSL) transport QDC channel at gate voltages of -1.3V and -0.3V, respectively. The voltage range can be adjusted by varying the work function used.

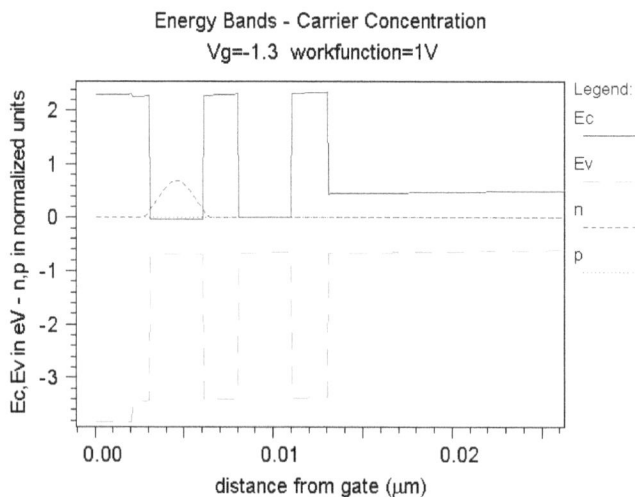

Fig. 5(a). Electron wavefunction at V_g = -1.3 V in GeO_x-Ge upper quantum dot layer.

Energy Bands - Carrier Concentration
Vg=-0.3 workfunction=1V

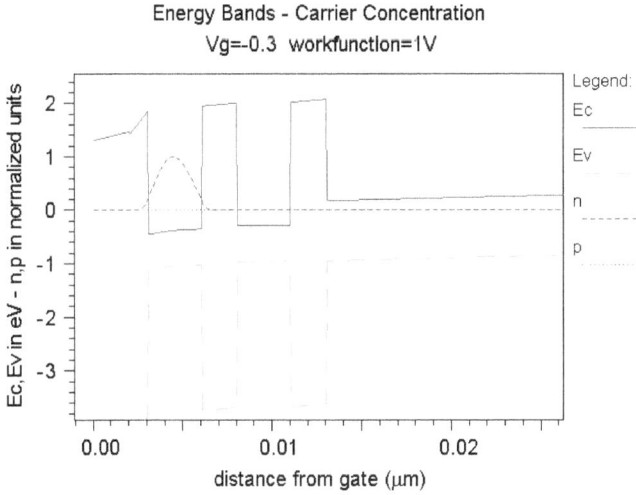

Fig. 5(b). Electron wavefunction at V$_g$ = -0.3 V in GeO$_x$-Ge upper quantum dot layer.

As the gate voltage is increased from -1.3V to -0.3V, the magnitude of charge in the upper quantum dot channel increases.

The parameters are shown in the two tables below. Thickness in micron (μm), electron affinity χ (eV), band gap E$_g$, effective electron and hole masses (m$_e$ and m$_h$), dielectric constant ε$_r$, and donor and acceptor doping levels (N$_d$ and N$_a$).

Table of parameters for GeO$_x$-Ge quantum dot channel FET.

Layer	Thick(μm)	χ(eV)	E$_g$(eV)	m$_e$	m$_h$	ε$_r$	N$_d$	N$_a$
1. HfAlO$_2$	0.0020	2.2	6.13	0.21	0.20	14.7	0.0e00	0.0e00
2. GeO$_x$	0.0010	2.25	5.70	0.16	0.16	4.4	0.0e00	0.0e00
3. Ge – QD 1	0.0030	4.55	0.67	0.08	0.28	16.0	0.0e00	0.0e00
4. GeO$_x$	0.0020	2.25	5.70	0.16	0.16	4.4	0.0e00	0.0e00
5. Ge – QD 2	0.0030	4.55	0.67	0.08	0.28	16.0	0.0e00	0.0e00
6. GeO$_x$	0.0020	2.25	5.70	0.16	0.16	4.4	0.0e00	0.0e00
7. Si	0.1000	4.15	1.12	0.19	0.49	11.9	0.0e00	1.0e16

Table of parameters for HfAlO$_2$ gate oxide (obtained from an average).

Layer	Thick(μm)	χ(eV)	E$_g$(eV)	m$_e$	m$_h$	ε$_r$	N$_d$	N$_a$
HfO$_2$	0.0000	2.4 (calc)	5.3-5.7	0.22	0.15	20.0	0.0e00	0.0e00
Al$_2$O$_3$	0.0000	1.95	6.95	0.2	0.25	9.34	0.0e00	0.0e00

The charge density in the transport channel is shown in Fig. 6. Here the charge density increases until the mini-band is filled and saturation occurs. When the gate voltage is increased along with drain voltage, second mini-band is made available and charge density increases. At a certain value of V_G, saturation is reached.

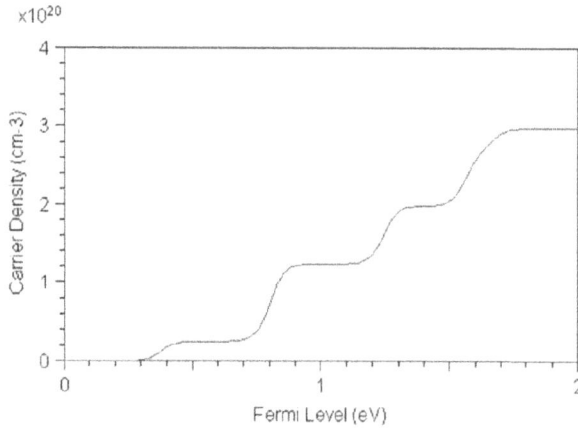

Fig. 6. Charge density plot as a funciton of Fermi level, related to gate voltage V_G.

Further increase in V_G and V_D is needed to open the third mini-energy band. Figure 7 shows the simulated I_D-V_D characteristics.

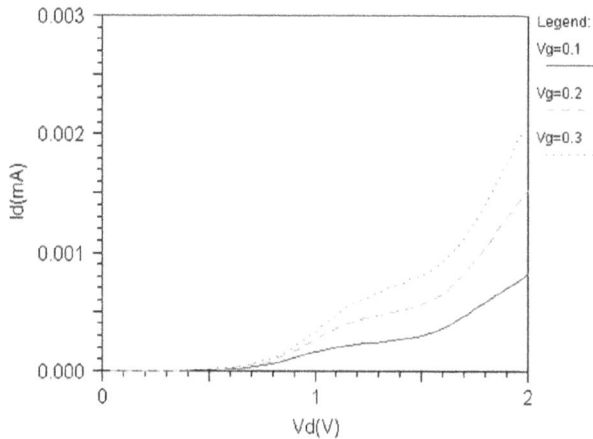

Fig. 7. Simulated I_D-V_D characteristics.

The drain current I_D is empirically expressed by Eq. (1) [3]. It depends on the number of mini-bands (i) and V_{DSj}, (signified by integer j). The number of mini-bands is determined by the value of V_{DSj} for a given V_G which determines the overall electron charge in the quantum dot channel. Therefore, the onset of various mini-bands can be simulated as if the device has various threshold voltages V$_{THi}$. The threshold shift ΔV$_{TH}$ depends on the transfer of charge to the QDs in the gate region in QDG-QDC FETs. Here, the quantum dot superlattice forms in quantum dot layers in the gate region [3].

$$I_D = \left(\frac{W}{L}\right) C_o'' \, \mu_n \left[\sum_{i=0}^{m} \sum_{j=0}^{n} \left\{ V_G - \left(V_{THi} + \Delta V_{THi}\right) - \frac{1}{2} V_{DSj} \right\} V_{DSj} \right] \tag{1}$$

5. Analog Behavioral M model

Analog behavioral model is also developed to use QDC-FET as a circuit element. The simulated characteristics using BSIM3 is shown in Fig. 8.

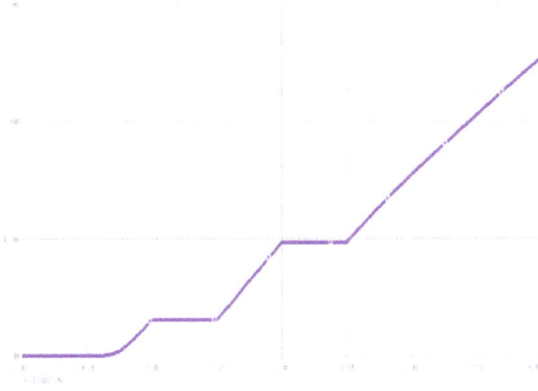

Fig. 8. I$_D$-V$_G$ characteristic.

6. Processing

The GeO$_x$-Ge QDC FET on poly-Si thin film, deposited on SOI substrate was fabricated using HfAlO$_2$ high-k gate dielectric over the GeO$_x$ cladding layer of quantum dot transport channel. A recessed region was created using Si$_3$N$_4$ masking layer. In addition, we deposted 75Å Si$_3$N$_4$ over p-doped poly-Si (1-5*10^{17}cm^{-3}) thin film. GeO$_x$-cladded Ge dots were prepared by a method reported elsewhere [4].

7. Conclusions

Quantum dot channel field effect transistors (QDC-FETs), using GeO$_x$-cladded Ge quantum dots on poly-Si thin films grown on silicon-on-insulator (SOI) substrates, have been successcully fabricated. HfAlO$_2$ high-k dielectric layers are used for the gate dielectric. QDC-FETs exhibit multi-state I-V characteristics. Multi-state characteristics (4-states and higher) enable 2-bit processing, and reduce FET count and power dissipation. QDC-FETs using cladded germanium quantum dots provide higher electron mobility than conventional poly-silicon FETs, and mobility values comparable to conventional FETs using single crystalline Si.

References

1. F. Jain, S. Karmakar, P.-Y. Chan, E. Suarez, M. Gogna, J. Chandy, and E. Heller, "Quantum Dot Channel (QDC) Field-Effect Transistors (FETs) Using II-VI Barrier Layers," *Journal of Electronic Materials*, June 23, (2012).
2. J. Chandy and F. Jain, *Proc. of International Symposium on Multiple Valued Logic*, pp. 186, (2008).
3. F. Jain, M. Lingalugari, J. Kondo, P. Mirdha, E. Suarez, J. Chandy, and E. Heller, "Quantum Dot Channel (QDC) FETs with Wraparound II-VI Gate Insulators: Numerical Simulations," *Journal of Electronic Materials*, 4, 11, (2016).
4. R. Velampati, El-Sayed Hasaneen, E.K. Heller, and Faquir C. Jain, "Floating Gate Nonvolatile Memory using Individually Cladded Monodispersed Quantum Dots," IEEE Transactions on Very Large Scale Integration (VLSI) Systems, 25, No. 5, May 19, (2017).

Quantum Dot Floating Gate Nonvolatile Random Access Memory Using Ge Quantum Dot Channel for Faster Erasing

Murali Lingalugari

GLOBALFOUNDRIES, Malta, NY 12020, USA

Evan Heller

Synopsis Inc., Ossining, NY 10562, USA
evankheller@gmail.com

Barath Parthasarathy, John Chandy, Faquir Jain*

Electrical and Computer Engineering, University of Connecticut,
371 Fairfield Way, Unit 4157, Storrs, CT 06269, USA
*faquir.jain@uconn.edu

This paper presents an approach to enhance floating gate quantum dot nonvolatile random access memory (QDNVRAM) cells in terms of higher-speed and lower-voltage Erase not possible with conventional floating gate nonvolatile memories. It is achieved by directly accessing the floating gate layer with a Ge quantum dot access channel via an additional drain (D2) during the Erase and/or Write operation. Quantum mechanical simulations in GeO_x-cladded Ge quantum dot layers functioning as the floating gate as well access channel to facilitate Erase and Write are presented. Experimental data on fabricated long channel nonvolatile random access memory cell with SiO_x-cladded Si dots is presented. Quantum simulations show lower voltage operation for GeO_x-cladded Ge QD floating gate than SiO_x-cladded Si dots. The Erase time is orders of magnitude faster than flash and is comparable to competing NVRAMs.

Keywords: Nonvolatile random access memory; quantum dot floating gate.

1. Introduction

Conventional flash memories use charge traps in the floating gate silicon nitride layer in Silicon-Oxide-Nitride-Oxide-Silicon (SONOS) structures. Tiwari *et al.* [1, 2] utilized nano-crystalline silicon dots as a floating gate which replaced the traps in the silicon nitride layer. However, Si quantum dots (QDs) were non-uniform in dot sizes as well as inter-dot spacing. Use of metallic nanocrystals [3-7] and the replacement of uncladded quantum dots with cladded quantum dots [8-13] has been reported. The cladded quantum dot (SiO_x cladded Si and GeO_x cladded Ge) floating gate layer reduces the charge leakage because the cladding acts as a barrier even if there are defects in the tunnel oxide. Flash memories

---

*Corresponding author.

suffer from longer erase times, which limits their usage as nonvolatile random access memories (NVRAMs), even in NOR architecture. In this paper, we present cladded Ge quantum dots as nonvolatile random access memories (QDNVRAMs) with improved performance both in terms of access time (sub-ns) and low-power operation.

1.1. *Ge QD-NVRAM structure*

Unlike recently reported Si QD NVRAMs [13], here the floating gate, as well as quantum dot access channel (QDAC) comprise of Ge QDs. Figure 1(a) shows the 3-D schematic of a quantum dot nonvolatile RAM with secondary drain D2 contacting the top two QD layers in the floating gate. The cell topology along with word line, bit line, and dedicated Erase line is shown in Fig. 1(b).

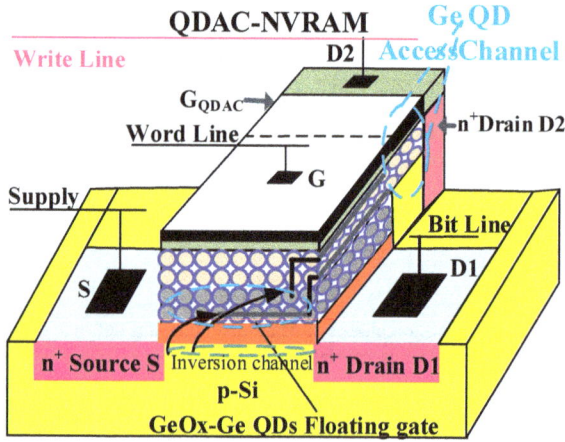

Fig. 1(a). The 3-D schematic of a quantum dot nonvolatile memory with four layers of self-assembled cladded Si quantum dots as the floating gate.

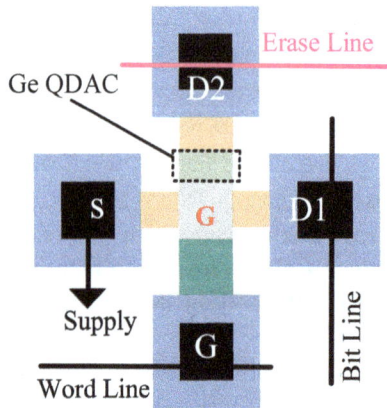

Fig. 1(b). Topology of a QDNVRAM cell showing drain D2 in relation to source S, drain D1, and gate G.

Our experimental results on SiO_x-Si cladded QD floating gate and QDAC demonstrate bit storage in long-channel ($10\mu m/14\mu m$) QDNVRAMs with ~4μs erasing. Our preliminary data on fabricated QDNVRAMs compares favorably to competing technologies [14-15] such as phase change memories (PCMs), magnetic RAMs (MRAMs), and resistive RAMs (RRAMs) in terms of erase speeds and power dissipation.

Four layers of SiO_x cladded Si QDs (~4nm Si core and ~1nm SiO_x cladding layer) are self-assembled over the tunnel oxide. The bottom two QD layers, in proximity of the tunnel oxide, form the floating gate while the top two quantum dot layers in the floating gate create a quantum dot access channel to access the stored electrons directly during the Erase operation through D2.

Cladded QDs with average particle size (APS) less than 6 nm are shown in the histogram of Fig. 2. Similar data is obtained for Ge quantum dots. During site-specific self-assembly of quantum dots, it is critical to maintain the optimum pH of the colloidal SiO_x-cladded Si quantum dot solution between 5 and 5.5 to ensure that the QDs are positively charged [17-18]. The value of pH for Ge quantum dot self-assembly is in the range of 3.25-4.25 [11]. In addition, the annealing temperature for Ge dots is lower than Si dots.

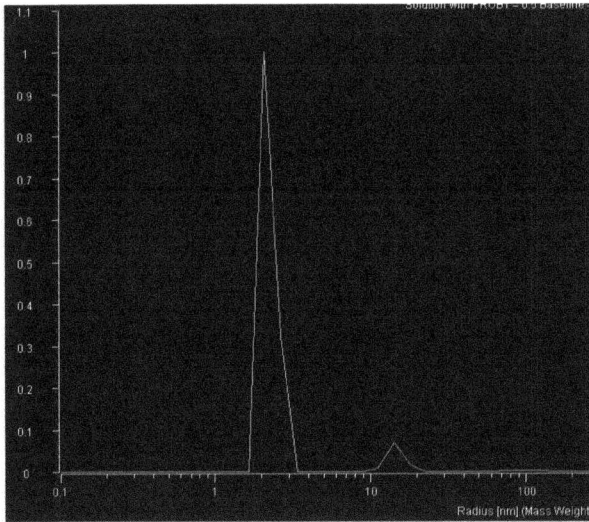

Fig. 2. DLS of SiO_x-Si QDs showing result an APS of less than 6nm with little variation.

The control gate dielectric consists of a ~7.5nm thick layer of silicon nitride or HfO_2 upon which a 100nm thick aluminum layer was deposited. The control gate dielectric thickness is quite smaller in the vicinity of drain D2 where the quantum dot access channel forms. The additional drain D2 and gate contacts are electrically isolated.

1.2. *Operation*

The Write mechanism of a QDNVRAM is similar to that of a conventional flash memory. The additional drain D2 of a QDNVRAM is not used and biased during the Write operation. The charges stored in the quantum dot floating gate layer of a QDNVRAM can be removed using conventional Erase mechanism via control and tunnel gate dielectric layers by biasing the gate and source. In addition, two high-speed erasing mechanisms are available by biasing secondary drain D2. Positive D2 bias removes the stored charges in the floating gate by attracting the electrons towards D2, whereas the negative D2 bias injects the electrons to the source by repelling the electrons away from D2. During the Write operation based on the applied gate and drain D1 voltages and duration, the electrons from the inversion channel tunnel to the bottom QD layer (adjacent to the tunnel oxide) near drain D1. As the control gate (V_G), drain D1 (V_{D1}) voltages and pulse durations increase further, the electrons starts filling the upper QD layers. This results in a change in the threshold voltage (V_{TH}) and manifests multi-bit storage. The transport of electrons within each QD layer is due to the formation of mini-energy bands, which forms quantum dot superlattice (QDSL) layers when the array of QDs are formed with thin cladding/barrier (SiO_x) layer.

2. Quantum Simulations-Two Layers of QDs in Floating Gate

2.1. *Si QD floating gate*

Figures 3(a) and 3(b) show the quantum mechanical simulations of the device with two SiO_x-cladded Si QD layers in the floating gate with a 7.5nm Si_3N_4 control dielectric layer. Figure 3(a) shows the formation of the inversion channel between the source and drain (D1) terminals at $V_G = 0.7V$. As V_G increases to 0.8V, the electrons tunnel to the bottom QD layer in the floating gate region as shown in Fig. 3(b).

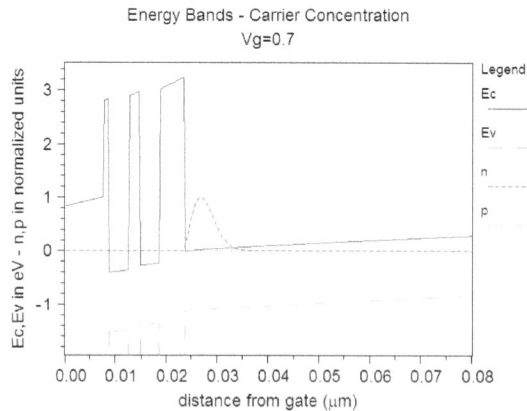

Fig. 3(a). Quantum simulations showing the formation of the electron inversion channel at $V_G = 0.7V$.

Energy Bands - Carrier Concentration
Vg=0.8

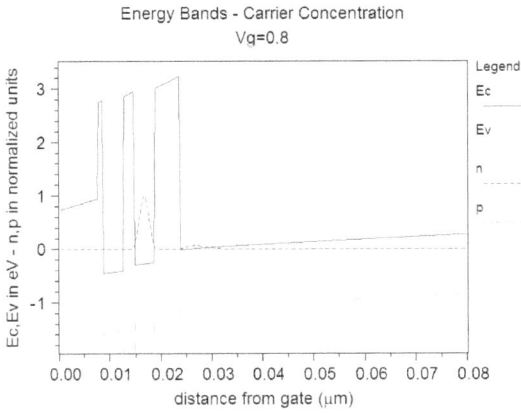

Fig. 3(b). Quantum simulations showing tunneling of electrons from the inversion channel to the bottom QD floating gate layer at $V_G = 0.8V$.

2.2. *Ge QD floating gate*

The tunneling of electrons in a Ge QDNVRAM device having 2 layers of GeO$_x$-cladded Ge QD layers is shown in Figs. 4. Figure 4(a) shows the formation of inversion layer at $V_g = 0.6V$. The electrons transfer from the inversion layer to first layer of Ge QDs in the floating gate when V_g is raised to 0.7V.

Energy Bands - Carrier Concentration
Vg=0.6, 1nm cladding

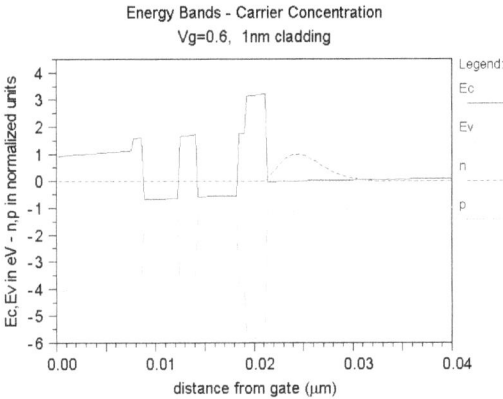

Fig. 4(a). Quantum simulations showing the formation of electron inversion channel at $V_G = 0.6V$.

Further increase in V_G to 0.8V results in transfer of some of the electrons to the second Ge QD layer [Fig. 4(c)], and at $V_G = 0.9V$ all electrons transfer to the second QD layer.

Energy Bands - Carrier Concentration
Vg=0.7, 1nm cladding

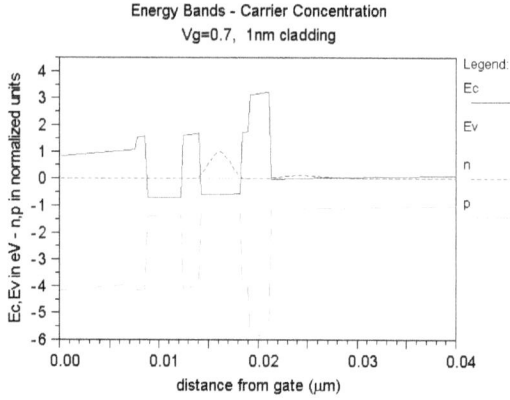

Fig. 4(b). Quantum simulations showing tunneling of electrons from the inversion channel to bottom QD layer at $V_G = 0.7$V.

Energy Bands - Carrier Concentration
Vg=0.8, 1nm cladding

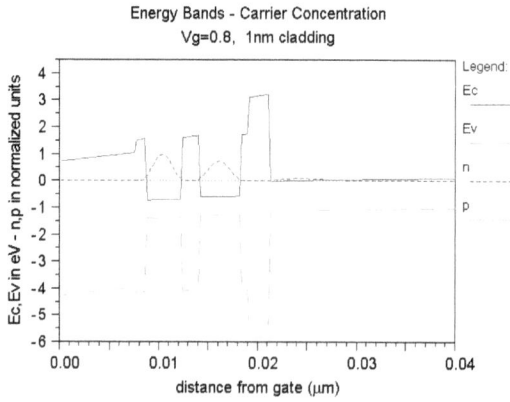

Fig. 4(c). Quantum simulations showing partial tunneling of electrons to the top Ge QD layer at $V_G = 0.8$V.

Energy Bands - Carrier Concentration
Vg=0.9, 1nm cladding

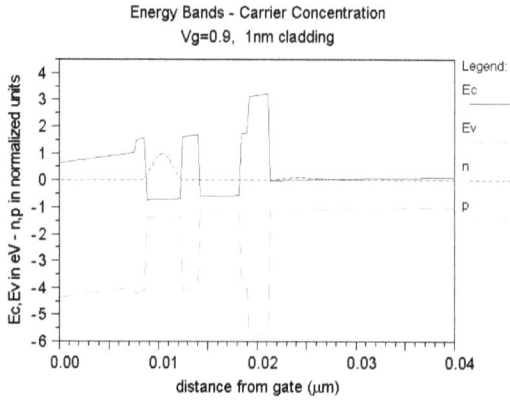

Fig. 4(d). Quantum simulations showing full tunneling of electrons to the top Ge QD layer at $V_G = 0.9$V.

The tunneling time as a function of cladding layer thickness is summarized in Fig. 4(e).

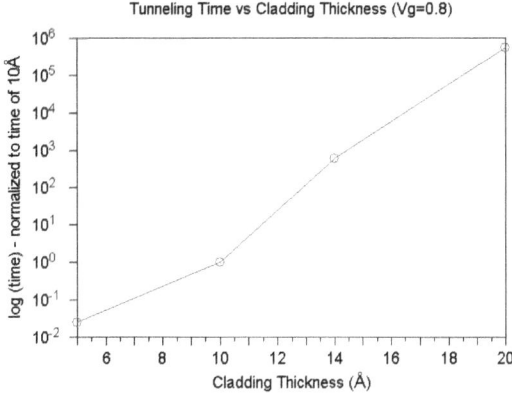

Fig. 4(e). Tunneling time as a function of GeO$_x$ cladding.

The charges stored in the QD layers increase the threshold voltage V$_{TH}$. The shift in the threshold voltage ΔV$_{TH}$ depends on the magnitude of gate voltage and pulse duration during Write #1. The threshold shift depends on location of electrons in the bottom, both, and top layer of the QDs.

Multiple bits could be written, therefore, by applying different values and pulse duration of V$_G$ pulse. The second QD layer in the floating gate gets electrons at a higher V$_G$ value. This transfer causes a different value of threshold voltage shift and can be treated as an additional stored bit. In addition, the duration of Write pulses can also provide another way of Write a multi-bit RAM cell. Comparing Fig. 3 and Fig. 4, it appears that Ge QDs memory cells could be written at lower voltages and more amenable for multi-bit operation. The threshold voltage shift (ΔV$_{TH}$) due to the charges present in the QDs can be expressed as equation (1) [13]:

$$\Delta V_{TH} = \frac{Q}{C_{CG-FG}} = \frac{\int_{o}^{t_w} j(t)Adt}{C_{CG-FG}} \tag{1}$$

Where, C$_{CG-FG}$ is the capacitance between the control gate and the QDs in the floating gate, A is the area of the floating gate, and j(t) is the current density while performing the Write or Read operation for a duration of t$_w$, which is given in equation (2) below

$$j(t) = q * n_{dot} * N_{QD} * P_{w \to d} \tag{2}$$

In equation (2), q is the charge of an electron, n$_{dot}$ is the number of electrons per dot (bound as well as the electrons trapped at the QDs interface), N$_{QD}$ is the density of QDs, and P$_{w \to d}$ is the tunneling rate of carriers from the channel (quantum well) to quantum dots. The amount of charges transferred to the QDs depends on the tunneling probability of the wavefunctions of the inversion channel (ψ_w) and the quantum dots (ψ_d). The charges in a QD layer can be calculated using electron-tunneling rate from the channel to the dots. The electron distribution in an inversion channel/quantum well can be calculated by solving the 1-D Schrödinger's and the Poisson's equations self-consistently [11-13].

3. Quantum Simulations for Four Layers of Ge Quantum Dots

The manifestation of electrons in Ge quantum dot access channel (two top layers of four layer QDNVRAM device) is shown in Figs. 5. Figure 5(a) shows the electrons transfer at V_G of 0.2V when the SiN control gate dielectric thickness of 7.5nm is used.

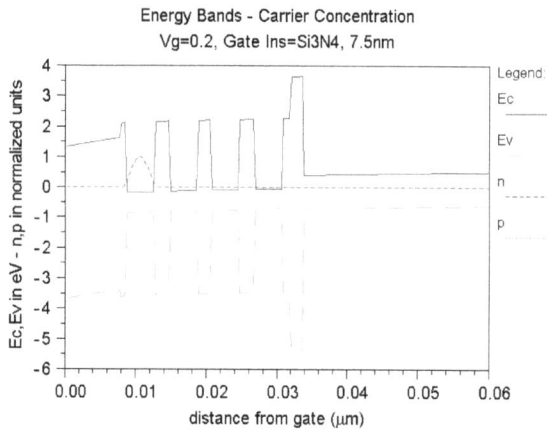

Fig. 5(a). Electron transfer at V_g = 0.2V (SiN control gate dielectric thickness is 7.5nm).

By contrast, the energy bands are slightly different when the SiN control gate dielectric thickness is 2.0 nm as shown in Fig. 5(b).

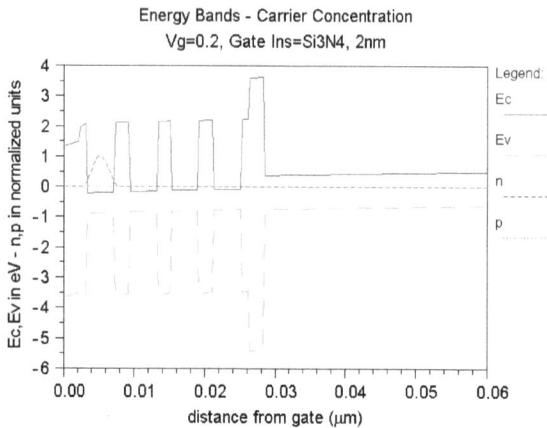

Fig. 5(b). Electrons transfer at V_g = 0.2V (Si$_3$N$_4$ control gate dielectric thickness is 2.0nm).

The parameters are tabulated in Table 1.

Table 1. Ge QD-NVM parameters.

Layer	Thick (μm)	χ (eV)	E_g (eV)	m_e/m_o	m_h/m_o	ε_r	N_d (cm-3)	N_a (cm-3)
Si₃N₄	0.0075 var*	2.7	5.0	0.16	0.16	7.5	0.0e00	0.0e00
GeOx clad	0.0010	2.25	5.70	0.16	0.16	4.4	0.0e00	0.0e00
GeQD core	0.0040	4.55	0.67	0.08	0.28	16.0	0.0e00	0.0e00
GeOx clad	0.0020	2.25	5.70	0.16	0.16	4.4	0.0e00	0.0e00
GeQD core	0.0040	4.55	0.67	0.08	0.28	16.0	0.0e00	0.0e00
GeOx clad	0.0020	2.25	5.70	0.16	0.16	4.4	0.0e00	0.0e00
GeQD core	0.0040	4.55	0.67	0.08	0.28	16.0	0.0e00	0.0e00
GeOx clad	0.0020	2.25	5.70	0.16	0.16	4.4	0.0e00	0.0e00
GeQD core	0.0040	4.55	0.67	0.08	0.28	16.0	0.0e00	0.0e00
GeOx clad	0.0010	2.25	5.70	0.16	0.16	4.4	0.0e00	0.0e00
SiOx	0.0020	0.9	9.0	0.5	0.5	3.9	0.0e00	0.0e00
Si	0.5000	4.15	1.12	0.19	0.49	11.9	0.0e00	1.0e16

*var = 7.5nm for Read, 2.0nm for Write/Erase **ϕ_b = -1.0eV

The simulation using HfO_2 control gate dielectric is shown in Fig. 5(c) and Fig. 5(d). Here, Si₃N₄ is replaced by HfO_2. The HfO_2 parameters used in simulation are: bandgap E_g=5.7eV, electron affinity χ=2.05eV, dielectric constant ε_r=13.7, and effective masses m_e=0.16 and m_h=0.5.

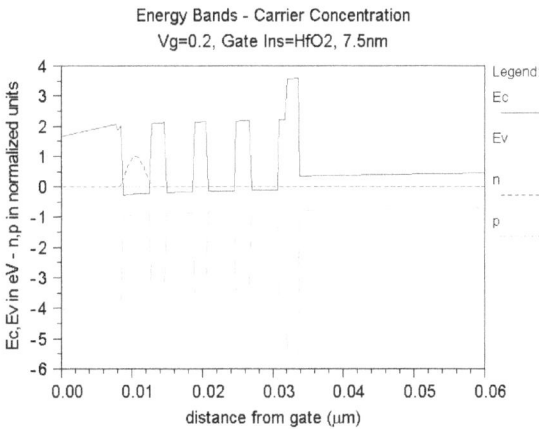

Fig. 5(c). Electrons transfer at V_g = 0.2V (HfO_2 control gate dielectric thickness is 7.5nm).

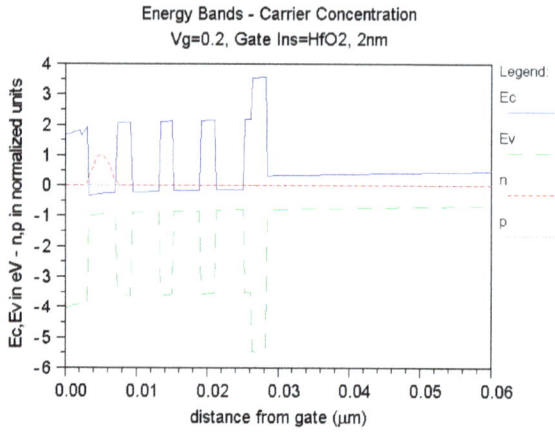

Fig. 5(d). Electrons transfer at $V_g = 0.2V$ (HfO$_2$ control gate dielectric thickness 2.0nm).

4. Experimental

The experimental I$_D$-V$_G$ characteristics of a Si QD NVRAM are shown in Fig. 6. Here, the W/L ratio of gate is 10μm/14μm, source and drain D1 extensions are 6μm x 20μm, and drain D2 extension is 14/20.5μm. The source and drain D1 regions (blue) are 50μm x 60μm, and drain D2 (blue) is 30μm x 45μm. The source and drain D1 contacts (black) are 30μm x 30μm, and drain D2 (black) contact 20μm x 35μm. Si QD NVRAM data shows preliminary work while Ge QDNVRAM experiments are in progress.

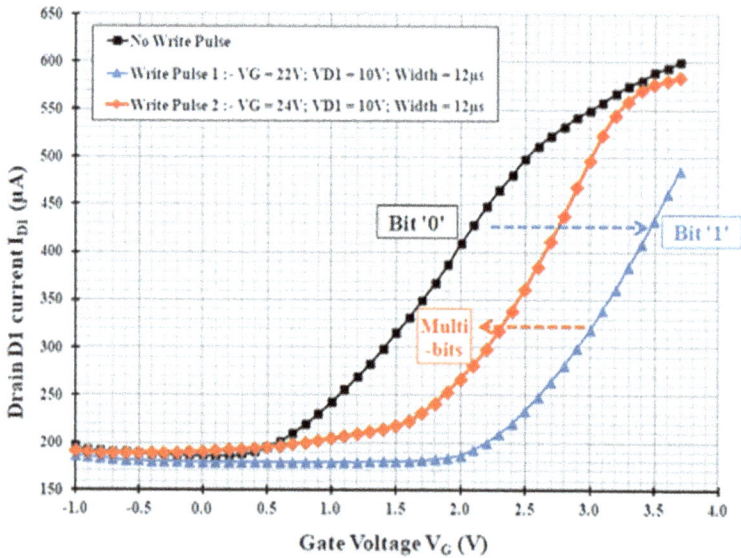

Fig. 6. Experimental I$_D$-V$_G$ Characteristics showing multi-bit.

When the applied V_G is greater than the V_{TH} of the device it results in the formation of an inversion channel. The black line (squares) represents no stored charge in the floating gate Bit '0'. When Write Pulse 1 is applied, the electrons from the inversion channel tunnel to the bottom Si QD layer, which is closer to the tunnel oxide, schematically presented in Fig. 3. Those stored carriers shift the V_{TH} of the device shown as the red line (circles) characteristic and the voltage shift is highlighted with the red arrow in Fig. 6.

5. Conclusions

In summary, quantum mechanical simulations in GeO_x-cladded Ge quantum dot layers functioning as the floating gate as well access channel, facilitating low-power high-speed (due to enhance mobility in Ge) Erase and Write, are presented. Quantum simulations clearly show lower voltage operation for GeO_x-cladded Ge QD floating gate than SiO_x-cladded Si dots. A combination of Ge and Si dots, forming first two layers of the floating gate, is also envisioned based on our prior work of SiGe QDG FETs.

Experimental data on fabricated long channel nonvolatile random access memory cell with SiO_x-cladded Si dots is presented. Figure 6 demonstrated in Si QD NVRAMs high-speed Erase in the range of 4-12µs for a long-channel (W/L ratio = 10µm/14µm) quantum dot floating gate nonvolatile random access memory at voltages much smaller than those used for the Write pulses. The Erase time is orders of magnitude faster than flash and is comparable to competing NVRAMs.

Acknowledgements

The authors gratefully acknowledge Dr. T.P. Ma and his laboratory personnel at Yale University for their assistance in memory testing. Authors also would like to acknowledge Dr. Ronald LaComb for his assistance in the device fabrication and the assistance of CNS facilities at Harvard University.

References

1. S. Tiwari, F. Rana, H. Hanafi, A. Hartstein, E. F. Crabbé, and K. Chan, Appl. Phys. Lett., **68**, 10, 1377 (1994).
2. S. Tiwari, F. Rana, K. Chan, L. Shi, and H. Hanafi, Appl. Phys. Lett., **69**, 1232 (1996).
3. M. Takata, S. Kondoh, T. Sakaguchi, H. Choi, J. C. Shim, H. Kurino, and M. Koyanagi, Tech. Dig. - Int. Electron Devices Meet. **2003**, 553 (2003).
4. C. Lee, A. Gorur-Seetharam, and E. C. Kan, Tech. Dig. - Int. Electron Devices Meet. **2003**, 557 (2003).
5. Z. Liu, C. Lee, V. Narayanan, G. Pei, and E. C. Kan, IEEE Trans. Electron Devices, **49**, 1606 (2002).
6. M. Kanoun, A. Souifi, T. Baron, and F. Mazen, Appl. Phys. Lett., **84**, 5079 (2004).
7. J. J. Lee, Y. Harada, J. W. Pyun, and D.-L. Kwong, Appl. Phys. Lett., **86**, 103505 (2005).
8. F. Jain, E. Suarez, M. Gogna, F. Alamoody, D. Butkiewicus, R. Hohner, T. Liaskas, S. Karmakar, P.-Y. Chan, B. Miller, J. Chandy, and E. Heller, J. Electron. Mater., **38**, 8, 1574 (2009).

9. E. Suarez, M. Gogna, F. Al-Amoody, S. Karmakar, J. Ayers, E. Heller, and F. Jain, J. Electron. Mater., **39**, 7, 903 (2010).

10. M. Gogna, E. Suarez, P.-Y. Chan, F. Al-Amoody, S. Karmakar, and F. Jain, J. Electron. Mater., **40**, 8, 1769 (2011).

11. M. Lingalugari, K. Baskar, P-Y. Chan, P. Dufilie, E. Suarez, J. Chandy, E. Heller, and F. C. Jain, J. of Electron. Mater., **42**, 11, 3156 (2013).

12. M. Lingalugari, P-Y. Chan, E. Heller, and F. Jain, Int. J. High Speed Electron. Syst., **24**, 03n04 (2015).

13. M. Lingalugari, P-Y. Chan, E. Heller, J. Chandy, and F. C. Jain, Electronics Letters, 54, No. 1, pp. 36-37, January 2018.

14. H.-S. P. Wong, S. Raoux, S. Kim, J. Liang, J. P. Reifenberg, B. Rajendran, M. Asheghi, and K. E. Goodson, Proceedings of the IEEE, Vol. 98, No. 12, pp. 2201-2227, (2010).

15. J. S. Meena, S. M. Sze, U. Chand, and T-Y. Tseng, Nanoscale Res. Lett., 9, p. 526, (2014).

16. R. Velampati, E.-S. Hasaneen, E. Heller, *et al.*, "Floating gate nonvolatile memory using individually cladded mono-dispersed quantum dots", *Trans. Very Large Scale Integer.*, 2017, **25**, pp. 1774-1781.

17. T. Phely-Bobin, D. Chattopadhyay, and F. Papadimitrakopoulos, Chem. Mater., **14**, 3, 1030 (2002).

18. F. Jain and F. Papadimitrakopoulos, U.S. Patent 7,368,370 (2008).

19. S. Chuang and N. Holonyak, Appl. Phys. Lett., **80**, 7, 1270 (2002).

Design of an Inductorless Power Converter with Maximizing Power Extraction for Energy Harvesting

Ridvan Umaz*

*University of Connecticut, Electrical & Computer Engineering,
Storrs, CT 06002, USA*
ridvan.umaz@uconn.edu

Lei Wang

*University of Connecticut, Electrical & Computer Engineering,
Storrs, CT 06002, USA*
lei.3.wang@uconn.edu

An inductorless power converter for low-power energy harvesting is presented. The power converter for energy harvesting is employed to maximize power extraction from energy sources. The power converter is based on a capacitive boost converter which is divided into two stages; a number of first-stage in parallel and shared-stage. The first-stage maximizes power extraction from the energy source while the shared-stage operates as a conventional charge pump. For not only low-power energy source but also high-power energy source, the maximum power extraction is targeted by the proposed converter. The extracted power from energy sources enhances by range from 117% to 161% over the conventional design. The output current of the proposed converter with three first-stages is improved by 183% over conventional converter. The peak efficiencies achieved with three and one first-stage are 53.3% and 38.5% for the proposed and the conventional converters, respectively. The peak end-to-end efficiency is enhanced by 198% as compared to the conventional converter. The proposed inductorless power converter has been implemented on a 0.13 μm CMOS process.

Keywords: Energy harvesting; maximum power point; charge pump; microbial fuel cell; solar cell; piezoelectronics.

1. Introduction

Energy harvesting utilizes the surrounding environmental energies to provide power to a wide set of applications, such as wireless sensor networks and remote devices [1, 2]. Energy sources including microbial fuel cells [3], piezoelectronics [4], photovoltaic cells [5], RF [6] and thermal energy [7] are serviceable for harvesters to utilize at various loads.

However, there are three main challenges in energy harvesting (i.e. specially low ambient power level) to power up the loads. First, the available energy harvested from the environment is extremely constrained by either ambient conditions or some design

---

*Corresponding author.

parameters. Energy sources typically generate low voltage and power at their outputs. For example, for photovoltaic cells, light intensity varies widely depending on locations and illuminance can range from 10s of lux at twilight or dim indoor conditions to 100Ks lux under direct sunlight. They can produce power from 5 W to 10mW [8]. Another example, microbial fuel cells exploit bio electrochemical reactions to generate power in the range of 10 W to 2mW, which highly depends on electrode size, installation distance [9, 10] and thermodynamic limitations [11].

Second, since energy harvesting starts with low ambient voltage and power levels, the generated voltages from energy sources are usually low, ranging from 10s of millivolts to 100s of millivolts. However, these voltages are not sufficient for electronic devices (e.g. sensors) to operate. In addition, energy sources are not suitable to directly power electronic devices, because the voltages at the outputs of energy sources can vary considerably during operation. Therefore, designing a low input voltage (e.g. less than 1V) power converter is crucial to up-convert the low voltage to a level usable by the load. There are two types of up-converters: an inductive type and a capacitive (i.e. inductorless) type. The inductive type converters are not fully-integrated due to external bulky components, and thus they are not preferable in many applications with size constraints. The capacitive converters can be an alternative solution. Some previous works have studied the inductorless power converter [12, 13, 14, 15] but they use general-purpose charge pumps implemented by either varying the number of stages on different load conditions or modulating switching frequency. These works usually require complex and power-hungry peripheral control circuits, thereby compromising low-power operation.

Finally, the harvesters need to be constructed efficiently to approach maximum power extraction. The maximum power available from the energy sources can be obtained by adjusting system parameters (e.g., number of stages, switching frequency). However, adjusting these system parameters need to either get help of DSP or CPU, or implement some power-consuming control logic circuits. One example, two different microbial fuel cells used as energy sources [12] generate the maximum power of 1.6mW and 11.2μW, and a maximum power extraction circuit is implemented to achieve this. However, the peak power dissipation of the circuit is 36.4μW, which is much higher than the low MFC power out. Therefore, this maximum power extraction circuit is not well suited for all energy sources. For low energy sources, maximum power extraction circuits should be ultra low power. Also, these circuits should be designed for a variety of power ranges in energy sources.

To overcome the above-mentioned challenges, this paper develops an efficient inductorless power converter for renewable energy sources to maximize the power extraction without introducing additional power overhead and circuit complexity. The proposed power converter employs a hybrid charge pump circuit that is divided into two stages. The first stage utilizes a number of pump units connected in parallel to maximize the power extraction from energy sources. The second stage is a shared-stage that acts as a conventional charge pump to step up the output voltage at the required level by the load. The proposed converter is achieved simply maximum power out from low-power energy sources owing to multiple first-stage instead of need of external sources i.e. battery and complex circuits. In comparison with conventional converter i.e. one first-stage, the proposed converter with three first-stage demonstrates some distinct advantages: (1) the maximize power extraction from energy sources is improved by range from 117% to 161% over the conventional one; (2) 183% more output current at the converter is obtained.

The rest of the paper is organized as follows. Section 2 describes the system architecture including the conventional power converter and the proposed one. Section 3 discusses how to achieve maximum power extraction. In Section 4, the design method and optimization of the proposed power converter are developed. Section 5 presents the detailed design of the proposed inductorless power converter circuit. Section 6 evaluates the simulation results, and conclusion is drawn in Section 7.

2. System Architecture

2.1. *Existing architectures*

Conventional energy harvesting circuits consist of an inductorless DC-DC converter which is based on capacitor switching including a charge pump and its driver circuit [12, 13, 14], as shown in Fig. 1(a). The power converter is composed of a number of sequential charge stages connected in series. Two complementary non-overlapped signals (CLK and CLKB) are generated by the driver circuit to drive the charge pump, charge the capacitor and switch on/off transistors in order to transfer the stored charges in the present stage to the following stage at the pump. Efforts to maximize power extraction from energy sources

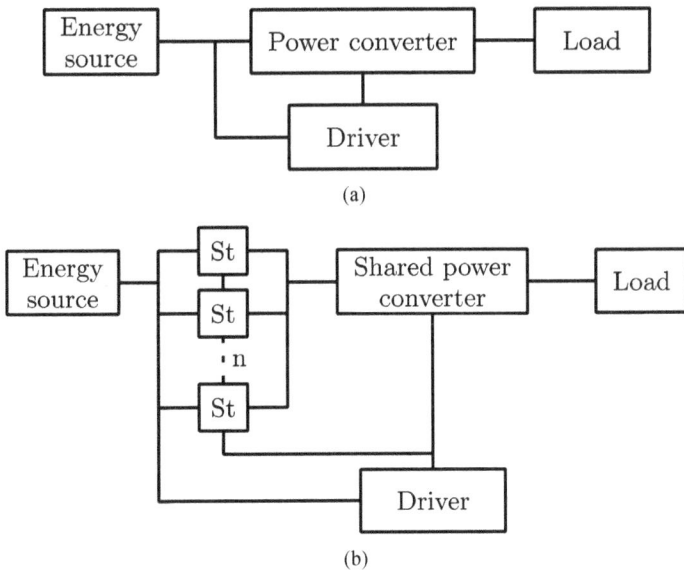

(a)

(b)

Fig. 1. (a) Conventional architecture with an inductorless power converter. (b) Proposed inductorless power converter.

(e.g., microbial fuel cells, photovoltaic and thermoelectric harvesters) with capacitive-based converters depend upon matching the internal impedance of the energy source with the converter, i.e., so-called impedance matching method.

The output impedance of a capacitive based converter is given by [16]

$$Req = \frac{N}{fC} \tag{1}$$

Where N is the number of stages, f is the switching frequency, and C is the pumping capacitor. It is obviously that the impedance can be adjusted by setting either the number of stages [13] or the switching frequency [12]. However, adjusting these parameters requires complex circuit implementations with power-hungry control logic. In addition, low-power output from the energy source does not provide sufficient power to adjust its impedance without using some external power supplies, and this affects the normal converter operation. There should be a more convenient way to maximize the power extraction from energy sources instead of using more complex circuit implementations.

2.2. *Proposed energy harvesting circuit*

This paper presents a more effective energy harvesting circuit as shown in Fig. 1(b). The proposed power converter consists of a power converter core and a driver circuit. The core is divided into two stages. The first stage includes a number of pumping stages connected in parallel. These pumping stages have the same stage circuit (St) whose outputs are combined at the same node. The merged node at the outputs of the pumping stages provides the input voltage for following stage (i.e., shared-stage).

The second stage is the shared-stage, which includes a number of sequential pumping stages connected in series that operate like a conventional charge pump. The driver circuit gets its supply voltage from the energy source. Although the driver circuit is similar to the conventional converter one, it has the capability to drive both the first-stage and the shared-stage, instead of employing separately drivers for each.

The output voltage of the proposed charge pump with N stages is expressed as

$$Vout = Vc + (N - 1) \times \Delta V = Vc + \sum_{i=1}^{N-1} \left(VL - \frac{Iout, i}{Ci \times f} \right) \tag{2}$$

Where ΔV is the voltage fluctuation at each pumping node. V_L is the clock supply voltage (input voltage of the energy source), V_c is the output voltage of the first-stage, C_i is the pumping capacitance at the ith stage, f is the clock frequency and I_{out} is the output current.

Note that the impedance matching method is applied to the proposed power converter as well. However, the impedance of the conventional converter expressed in (1) is not applicable to the proposed converter. New variables should be incorporated into 1 and the proposed power converter output impedance is given by

$$\text{Reqpro}=(\frac{N}{f\times C})\times\frac{1}{k} \tag{3}$$

Where k is a factor that is associated with the number of the pumping stages at the first-stage connected in parallel. The factor k should be less than the number of the pumping stages at the first-stage (k < n). This is mainly due to the fact that the number of switching transistors and the pumping capacitors at the shared-stage is smaller than n conventional power converters connected in parallel including n driver circuits.

As the number of pumping stages at the first-stage increases, the current drawn from the energy source also increases while the voltage at the output of the energy source decreases. Thus, the power extracted from the source will approach the maximum power point. Thus, the proposed converter improves the power extraction from the energy source as compare to the single first-stage implementation (i.e., conventional one), thereby transferring more extracted power to the shared-stage.

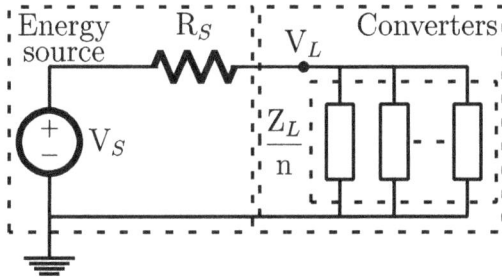

Fig. 2. Thevenin electrical equivalent circuit for energy sources (e.g. solar, MFCs, TEGs).

3. Maximum Power Extraction

Each energy source has different electrical models to represent their internal equivalent circuits. While solar cells can be modeled as a current source in parallel with a diode, MFCs and TEGs can be modeled as a voltage source in series with a resistor. However, all energy sources can be generally modeled as a Thevenin voltage in series with the Thevenin equivalent impedance, which could include resistance, capacitance and inductance. The Thevenin equivalent circuit for energy sources (e.g. solar, MFCs and TEGs) with N power converters connected in parallel and sharing a common driver circuits is shown in Fig. 2. Note that the proposed converter including N pumping stages connected in parallel at the first-stage acts as N power converter connected in parallel. Maximum power extraction from an energy source is obtained once the input impedance of the connected components (e.g. boost converters, resistive loads, charge pumps) interfacing with the source is viewed to be equal to the source internal impedance. This is referred to the impedance matching theory. The maximum power can be obtained as

$$Pmax = \frac{Vs}{4 \times Rs} \qquad (4)$$

In Fig. 2, the power delivered to the N power converters connected in parallel can be described as

$$PL=(\frac{Vs}{Rs+\frac{ZL}{n}})^2 \times \frac{ZL}{n} \qquad (5)$$

The matching efficiency η_Z can be viewed as the ratio of P_L to P_{max}, given by

$$\eta Z = \frac{PL}{Pmax} = \frac{4}{\frac{ZL}{n \times Rs} + \frac{n \times Rs}{ZL} + 2} \qquad (6)$$

Figure 3 shows the matching efficiency as a function of $Z_L = R_S$ under various values of N. More than 90% of the available power can be extracted from the energy sources once the endurable impedance mismatches ranges from -48% to +93% for the conventional converter. The impedance mismatch range increases proportionally with the number of the converters connected in parallel. For instance, once two converters are connected in parallel, the impedance mismatches ranges from -2 48 (-96%) to +2 93 (+186%) for more than 90% of the matching efficiency. This shows the number N of converters has a significant error tolerance over the conventional design. Therefore, the output power remains very close to the maximum power point (MPP) even when a large impedance mismatch occurs in the proposed converter design. This is a simple and self-starting (i.e., no need for extra power supplies) method to maximize the power extraction, instead of employing more complex power-hungry peripheral circuits as in conventional converter designs.

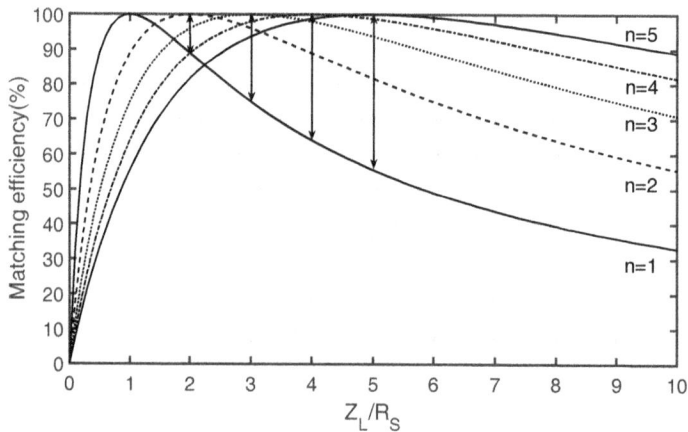

Fig. 3. Matching efficiency η_Z as a function of Z_L/R_S.

4. Design Optimization

Maximizing power extraction from energy sources and achieving a high power conversion efficiency at the end of power converters are crucial, in particular when low power sources are used as the energy source. To maximize power out and acquire high efficiency, some design parameters (e.g. current consumption, capacitor values, transistor sizes, etc.) need to be optimized and selected appropriately. Some previous works [17, 18] discussed a general strategy for charge pump design optimization. However, this method is insufficient to the proposed converter. A new design and optimization method is required in order to achieve the best efficiency.

In Section 2.2, the output voltage is given as a function of the first-stage output voltage V_c as 2. For simplicity, the first-stage pump capacitance is not considered. However, design optimization throughout the proposed converter should include the first-stage pump parameters with shared-stage ones.

The output voltage of the proposed converter with N stages can be expressed as

$$Vout = (N+1) \times VL - \left[\frac{Iout}{f \times Cf \times n} + \frac{Iout}{f \times Cs} \times (N-1) \right] \tag{7}$$

Where I_{out} is output current, f is the switching frequency, n is the number of the pumping stages at the first-stage, C_f and C_s are pump capacitances of the first-stage and shared-stage, respectively.

Note that there are two terms in (7). The first one refers to the pump output voltage ((N+1) V_L) in the case of a pure capacitive load, and the second one presents voltage loss in the case of a current load. The second term also has two components: the first one reflects the effect of the first stage on the output of the power converter, and the second one shows the effect of the shared stage which is voltage drop on the charge pump output once the load is connected.

The number of the pumping stages at the shared-stage is determined by (2) and can be given by

$$M = \frac{Vout}{Vc} - 1 \tag{8}$$

The first-stage pumps are counted as one pumping stage without considering the number of pumping stages in parallel, due to the same voltage V_c at the output of the first stage. The total number of pumping stages at the proposed converter is thus given by

$$N = M + 1 \tag{9}$$

The number of the pumping stage at the first stage depends on the particular energy source being used.

In order to fully transfer the stored charge from a capacitor (e.g. C_f) at the first stage to a capacitor (e.g. C_s) at the shared stage, the capacitance of the first stage should be much smaller than the shared-stage one. Otherwise, the remaining charges at the first stage will reduce the pumping capability. This can be set as

$$Cf \times n \leq Cs \qquad (10)$$

Where n is the number of the pumping stages at the first stage.

Substituting (10) into (7), we can derive the capacitance of a stage in (7) as

$$Cf = \frac{Iout \times (N-1)}{f \times n[(N+1) \times VDD - Vout]} \qquad (11)$$

The transistors also need to be sized accordingly in order to support the efficient charge transfer from the first stage to the shared stage.

The input and output capacitors should also be considered. The input capacitor C_{in} is critical since energy sources are low power. The capacitor accumulates the harvested energy so as to provide supply voltage for the driver circuits as well the input voltage for the converter.

5. Circuit Implementation

The block diagram of the proposed power converter is shown in Fig. 4(a). The converter consists of three first-stage and two shared-stage pumps, and two non-overlapped clock signals. Due to the inherent low power at the output of the energy source, a temporary energy buffer (e.g. C_{in}) is used to accumulate the harvested energy so as to provide supply voltage to the converter. Input capacitor C_{in} of 100nF and output capacitor C_{out} of 1nF were used.

A detailed schematic of the stage (St) is shown in Fig. 4(b). The St consists of a voltage doubler, a dual-series PMOS switch (M_{p1} and M_{p2}), and two non-overlapped complementary signals (CLK and CLKB). The voltage doubler includes a cross-coupled

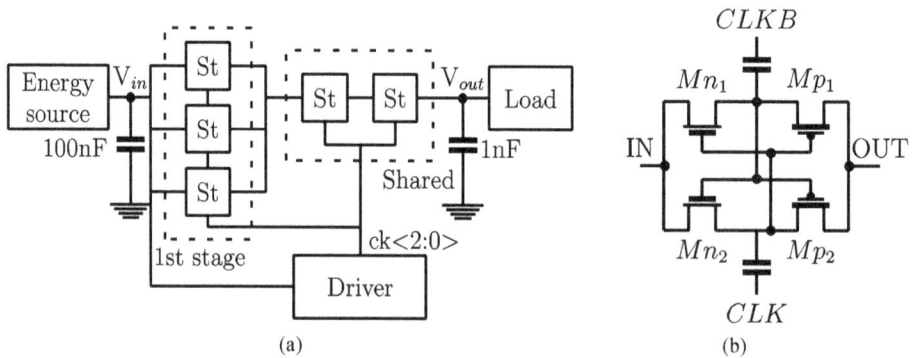

Fig. 4. (a) Block diagram of the proposed power converter. (b) Circuit implementation of the charge pump stage (St).

NMOS pair (M_{n1} and M_{n2}), and two pumping capacitors. The voltage doubler allows the capacitors' nodes to swing between input voltage (IN) and output voltage (OUT), which is equal to 2 IN. The same stage circuit (St) is used for both first stage and shared stages, expect that some parameters (e.g. pumping capacitor, transistor size) are different. For evaluation purpose, the pumping capacitor of 20pF for the first stage and of 60pF for the shared stage was used. The transistors are sized accordingly to meet the requirement on output voltage.

The block diagram of the driver circuit is shown in Fig. 5. The driver consists of a n-stage ring oscillator and a non-overlapping clock generator. The n-stage ring oscillator, whose supply voltage is obtained from the energy source, generates a clock for the non-overlapping clock generator to generate two non-overlapped complementary clocks (CLK and CLKB). The clock generator includes two NAND gates, a transmission gate, and a couple of inverters. The switches at the St circuit are not turned on simultaneously, which are controlled by the non-overlapping clocks. As a result, power efficiency degradation is kept low.

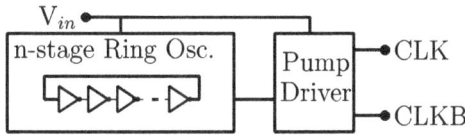

Fig. 5. Block diagram of the driver circuit.

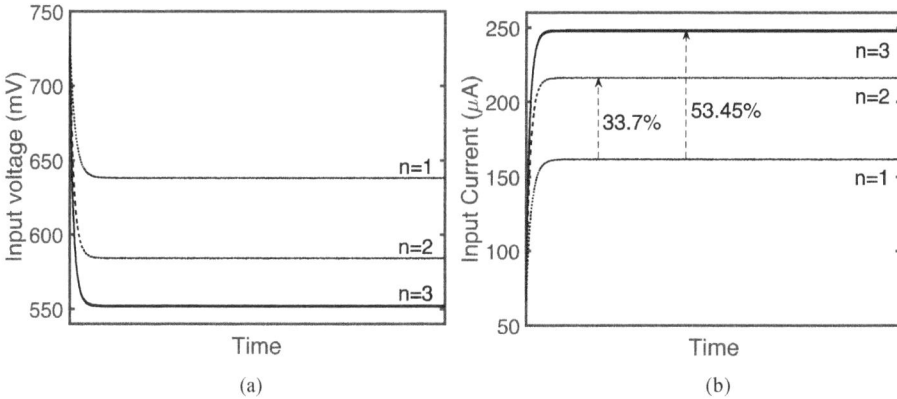

Fig. 6. (a) Input voltage of the proposed converter under varied number of first-stage. (b) Input current of the proposed converter under varied number of first-stage.

6. Results and Discussion

The proposed power converter was evaluated in a 0.13μm CMOS process. For simulations throughout this work, an 800 mV input voltage source in series with a 1k internal resistor was used to emulate the energy source.

In order to evaluate power extraction from the energy source, we vary the number of first stage in the proposed power converter. The voltages and currents at the input of the converter under different numbers of first stage units are monitored once a load resistor of 10k is connected, as shown in Fig. 6. The input voltage at the converter decreases with the increase in the number of the first stage, while the input current increases when adding more first stage units. The converter with three first stage units draw 53.45% more current from the energy source than the conventional converter with one first stage. Once one first stage unit is employed, the power of 103 W is extracted from the energy source. When three first stage units are used, the extracted power is 137 W, which is 33% more. Under varied loads, the improvement of extracted power with proposed converter (e.g. n = 3) over the conventional converter (e.g. n = 1) ranges from 17% to 61%. As a result, the proposed converter with three first stage units approaches very close to the MPP than the conventional converter.

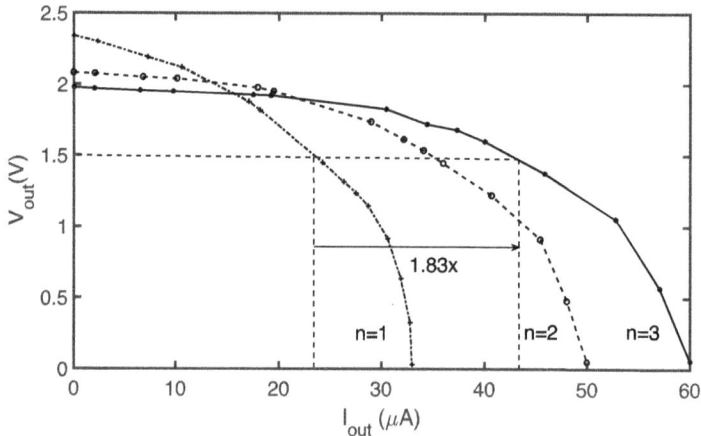

Fig. 7. The proposed converter output voltage as a function of its output current under varied number of first-stage.

Figure 7 shows the proposed converter output voltage as a function of its output current under different numbers of first stage units. In the light load mode, although conventional converter (e.g. n = 1) has larger output voltage than the proposed one, their output currents are nearly close to each other. The reason varied output voltage in the light load mode is that the proposed converter extracts more power than the conventional one. Drawing more current from energy source is obtained as number of the first-stage is increased. Therefore, the input voltage of the conventional converter is higher than the proposed ones. However, in the heavy load mode, the proposed converter has better output voltage and current than the conventional one. The proposed converter improves the maximum output current by 183% as compared to the conventional converter. As mentioned, the proposed converter extracts more power from the energy source. As a result, the proposed converter shows better power output closer to the MPP than the conventional converter. Figure 8 shows the power efficiency as a function of output current under different numbers of first stage units. Although the conventional converter exhibits better efficiency than the proposed one at the light load mode, the conventional converter has worse efficiency than the proposed converter at the heavy load mode. Heavy load requires more power than low one. The proposed converter extracts more energy than the conventional one from energy source. In the heavy load mode, the conventional converter transfers less power to the output, so power degradation is increased. The maximum efficiencies obtained from the proposed and conventional converter are 53.3% and 38.5%, respectively.

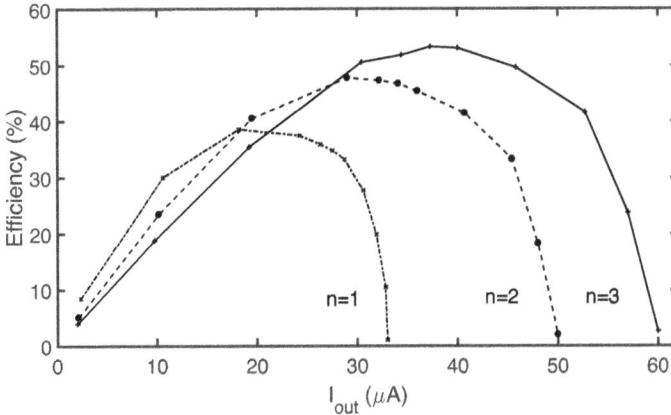

Fig. 8. The efficiency of the proposed converter under varied number of first stages.

However, the peak end-to-end efficiency (see (12)), which is the ratio between the power obtained at the load and the maximum power available from the energy source, varies for the proposed and the conventional converter, whose efficiencies are 39.6% and 20%, respectively. This indicates that the proposed converter improves the end-to-end efficiency by 198% as compared to the conventional converter. Trans erring more energy to the load is achieved at the proposed converter because more energy is extracted from energy source.

$$\eta end = \frac{Pin}{Pmax} \times \frac{Pout}{Pin} = \frac{Pout}{Pmax} \tag{12}$$

7. Conclusions

This paper presents an inductorless power converter for energy harvesting. As compared to conventional capacitive based power converter, the proposed power converter is divided into two parts; first-stage and shared-stage. First-stage is the maximum power extraction stage by connecting a number of first-stage in parallel. First-stage achieves maximizing power extraction without increasing power dissipation and circuit complexity except increasing area. Shared-stage operates as conventional charge pump to step-up the merged output voltage of a number of first-stage to a usable level by the application. Maximum power extraction is analyzed. Design methodology and optimization are discussed and incorporated into circuit implementation. The extracted power from energy sources enhances by range from 117% to 161% over the conventional design. The proposed inductorless power converter improves the end-to-end efficiency by 198% as compared to the conventional converter. The proposed power converter is useful not only for low energy sources but also for high energy sources. Note that the number of first-stage units in this work is manually chosen for the sake of demonstrating the improvement. Future work is being directed towards further enhancing the tracking efficiency based on automatically adjusting number of first-stage through circuit design implementation and the IC chip implementation of the proposed converter.

References

1. G. Guoxian, R. Umaz, U. Karra, L. Baikun, and W. Lei, "A biomass based marine sediment energy harvesting system," Proc. IEEE International Symposium on Low Power Electronics and Design (ISLPED), August 2013, pp. 359-364.
2. U. Karra, E. Muto, R. Umaz, M. Kolln, C. Santoro, L. Wang and B. Li, "Performance evaluation of activated carbon-based electrodes with novel power management system for long-term benthic microbial fuel cells," International Journal of Hydrogen Energy 39(12) (2014) 21847-21856.
3. R. Umaz, C. Garrett, F. Qian, B. Li, and L. Wang, "A power management system for multianode benthic microbial fuel cells," IEEE Transactions on Power Electronics 32(5) (2017) 3562-3570.
4. M. Dini, A. Romani, M. Filippi, and M. Tartagni, "A nanocurrent power management ic for low-voltage energy harvesting sources," IEEE Transactions on Power Electronics 31(6) (2016) 4292-4304.
5. K. W. R. Chew, Z. Sun, H. Tang, and L. Siek, "A 400nw single inductor dual-input- tri-output dc-dc buck-boost converter with maximum power point tracking for indoor photovoltaic energy harvesting," Proc. IEEE International Solid-State Circuits Conference Digest of Technical Papers, Feb 2013, pp. 68-69.
6. A. Mansano, S. Bagga, and W. Serdijn, "A high efficiency orthogonally switching passive charge pump rectifier for energy harvesters," IEEE Transactions on Circuits and Systems I: Regular Papers 60(7) (2013) 1959-1966.
7. A. Das, Y. Gao, and T. T. H. Kim, "A 220-mv power-on-reset based self starter with 2-nw quiescent power for thermoelectric energy harvesting systems," IEEE Transactions on Circuits and Systems I: Regular Papers, 64(1) (2017) 217-226.
8. Y. Qiu, C. V. Liempd, B. O. het Veld, P. G. Blanken, and C. V. Hoof, "5μW-to-10mW input power range inductive boost converter for indoor photovoltaic energy harvesting with integrated maximum power point tracking algorithm," Proc. IEEE International Solid-State Circuits Conference, Feb 2011, pp. 118120.
9. G. Huang, R. Umaz, U. Karra, B. Li, and L. Wang, "A power management integrated system for biomass-based marine sediment energy harvesting," International Journal of High Speed Electronics and Systems 23 (2014) 1-20.
10. U. Karra, G. Huang, R. Umaz, C. Tenaglier, L. Wang, and B. Li "Stability characterization and modeling of robust distributed benthic microbial fuel cell (DBMFC) system," Elsevier Bioresource Technology 144(9) (2013) 477-484.
11. J. D. Park and Z. Ren, "Hysteresis-controller-based energy harvesting scheme for microbial fuel cells with parallel operation capability," IEEE Transactions on Energy Conversion 27(3) (2012) 715-724.

12. S. Carreon-Bautista, C. Erbay, A. Han, and E. Sanchez-Sinencio, "An inductorless dc-dc converter for an energy aware power management unit aimed at microbial fuel cell arrays," IEEE Journal of Emerging and Selected Topics in Power Electronics 3(4) (2015) 1109-1121.

13. I. Doms, P. Merken, C. V. Hoof and R. P. Mertens, "Capacitive power management circuit for micropower thermoelectric generators with a 1.4μ a controller," IEEE Journal of Solid-State Circuits 44(10) (2009) 2824-2833.

14. Y. C. Shih and B. P. Otis, "An inductorless dc-dc converter for energy harvesting with a 1.2μW bandgap-referenced output controller," IEEE Transactions on Circuits and Systems II: Express Briefs 58(12) (2011) 832-836.

15. F. Qian, R. Umaz, Y. Gong, B. Li, and L. Wang, "Design of a shared-stage charge pump circuit for Multi-anode Microbial Fuel Cells," Proc. IEEE International Symposium on Circuits and Systems (ISCAS), May 2016, pp. 213-216.

16. J. F. Dickson, "On-chip high-voltage generation in mnos integrated circuits using an improved voltage multiplier technique," IEEE Journal of Solid-State Circuits 11(3) (1976) 374-378.

17. T. Tanzawa and T. Tanaka, "A dynamic analysis of the dickson charge pump circuit," IEEE Journal of Solid-State Circuits 32(8) (1997) 1231-1240.

18. G. Palumbo, D. Pappalardo, and M. Gaibotti, "Charge-pump circuits: power-consumption optimization," IEEE Transactions on Circuits and Systems I: Fundamental Theory and Applications 49(11) (2002) 1535-1542.

High-Speed Pulsed Fiber Ring Laser Using Photonic Crystal Fiber

Xiang Zhang[*]

Department of Physics, University of Connecticut, 2152 Hillside Road, U-3046, Storrs, CT 06269, USA
xiang.zhang@uconn.edu

Sunil Thapa

Department of Physics, University of Connecticut, 2152 Hillside Road, U-3046, Storrs, CT 06269, USA
sunil.thapa@uconn.edu

Niloy K. Dutta

Department of Physics, University of Connecticut, 2152 Hillside Road, U-3046, Storrs, CT 06269, USA
nkd@phys.uconn.edu

In this paper, we propose and experimentally demonstrate a mode locked fiber ring laser with the implementation of a photonic crystal fiber (PCF) to generate pulse train at high speed. This fiber ring laser combines rational harmonic mode locking based on a Lithium Niobate Mach-Zehnder modulator and nonlinear polarization rotation from a highly nonlinear PCF. By fine tuning of the modulation frequency and the polarization controllers in the cavity, a 30 GHz pulse train with pulse width 1.9 ps is generated. Without the PCF, the pulse width at 30 GHz from the rational harmonic mode locking is 5.8 ps. We also conduct numerical simulations of the pulse evolution, which shows good agreements of the experimental results.

Keywords: Fiber ring laser; rational harmonic mode locking; photonic crystal fiber; pulse compression.

1. Introduction

In future high-bit-rate fiber optic communication systems, stable high-speed pulse train with ultrashort pulse width will be very important in all-optical switching, encryption and signal processing [1, 2]. During the past decades, the generation of high speed short pulses has been studied extensively [3-9]. Among various platforms, active harmonic mode-locking has proven to be an effective way to generate high-repetition-rate pulses by incorporating an electro-optical intensity modulator, such as $LiNiO_3$ modulators or electric absorption modulators (EAM) inside the laser cavity. In particular, the implementation of rational harmonic mode-locking technique is able to overcome the modulator bandwidth

---

[*]Corresponding author.

limitation and further increase the repetition rate [6, 10]. However, with rational harmonic mode-locking, the temporal pulse duration is usually limited to several picoseconds. Furthermore, the amplitudes of pulse train suffer from severe fluctuations in the time domain, which is not good for fiber optic communication systems. Several researchers have carried out experiments to compress the optical temporal width or reduce the amplitude fluctuations. Ma *et al.* demonstrated a pulse width compression scheme by guiding the pulse train generated from a rational harmonic mode-locked fiber ring laser to pass through a nonlinear amplifying loop mirror twice [5]. Li *et al.* demonstrated theoretically and experimentally pulse-amplitude-equalization in 4th rational harmonic pulses based on nonlinear polarization rotation [11].

Recently, compact passively mode-locked fiber lasers have been successfully constructed to generate sub-picosecond pulses by using a saturable absorber (SA) such as semiconductor saturable absorber mirror (SSAM) [12], graphene [13, 14], nonlinear polarization rotation (NPR) technique [15-17] and nonlinear fiber loop mirror [18]. The NPR technique combined with a polarizer can induce an intensity depended loss in the cavity, and has been used to achieve ultrashort pulses in fiber lasers. Compared to actively mode-locked lasers, passively mode-locked fiber lasers have the advantage of generating pulse trains at ultrashort pulse width. In [15], Luo *et al.* constructed a L-band passively mode-locked fiber laser utilizing the NPR technique and generated pulses with full width at half maximum (FWHM) 458.7 fs. However, the pulse repetition rate is only at 8.6 MHz. Liu *et al.* reported the generation of a stable passive 23rd harmonic mode-locked pulse train at 230 MHz with a pulse width of 0.44 ps [17]. Despite the fact that those NPR based passively mode locked fiber lasers can produce ultrashort femtosecond pulses, they suffer from the drawback of low repetition rate (only at MHz level) with respect to the total cavity length, which limits their applications in high speed fiber optic communications. A possible solution would be to build a hybrid mode-locked scheme to combine this two mode-locking methods. Li *et al.* both numerically and experimentally demonstrated that by incorporating a charcoal nano-particle saturable absorber into the rational harmonic mode-locked laser cavity, they were able to generate a pulse train at high repetition rate (20 GHz) and short pulse width (~3.2 ps) [19].

Photonic crystal fiber (PCF), also called holey fiber, has a small core surrounded with air holes in the cladding. It's special properties due to the large refractive index difference between air and the core material make it a perfect candidate for nonlinear applications such as supercontinuum generation and pulse compressing [20, 21]. In this paper, we report a hybrid mode-locked erbium-doped fiber ring laser by combining the active rational harmonic mode-locking and the NPR based on a highly nonlinear PCF. A high speed 30 GHz pulse train with improved stability and narrower pulse width is generated. In this hybrid scheme, by carefully adjusting the polarization controllers the full width at half maximum (FWHM) of a 30 GHz pulse train is shortened to ~1.9 ps compared to ~5.8 ps with purely rational harmonic mode locking.

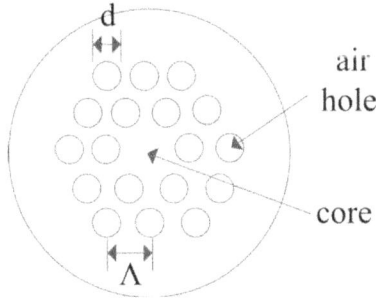

Fig. 1. Structure of a six-fold symmetric photonic crystal fiber. The core is surrounded by air holes with size d and spacing Λ.

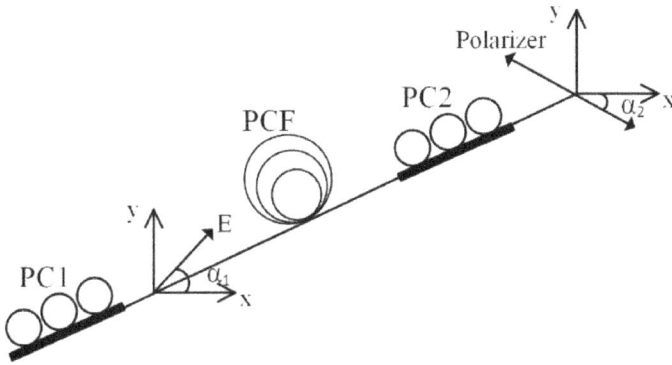

Fig. 2. Operation principle of NPR. E: electric field, x: fast axis of PCF, y: slow axis of PCF, PC: polarization controller.

2. Working Principle

Figure 1 gives the cross section of a photonic crystal fiber with a perfect six-fold symmetric core and cladding structure. In practice, due to the large glass-air index diffrenece, even a slight accidental distortion in the structure yeild a degree of birefringence [22]. Therefore, by deliberately distorting the core, extrem high values of birefringence can be achieved in a PCF. If combined with a polarizer, nonlinar polarization rotation will makes this configuration an aritificial saturable absorber. Therefore, the PC1-PCF-PC2-polarizer structure can introduce the intensity-dependent loss and the transmission principle is shown in Fig. 2. The PCF used here has dispersion parameter $\beta_2=-1.66$ ps^2/km, $\beta_3=-0.03$ ps^3/km and nonliearity $\gamma \sim 11$/W/km. The birefringence Δn is 3.5×10^{-5}. In Fig. 2, α_1 is the angle between the fast axis of the PCF and the polarization direction of the input signal before entering the PCF. E is the electric vector of input signal. α_2 is the angle between the fast axis of PCF and the polarization direction of the in line polarizer. The Kerr nonlinearity of the PCF can generate a rotation of polarization state, which depends on the pulse intensity. The transmission introduced by NPR can be expressed as [11]:

$$T = cos^2 \alpha_1 cos^2 \alpha_2 + sin^2 \alpha_1 sin^2 \alpha_2 + \tfrac{1}{2} sin 2\alpha_1 sin 2\alpha_2 cos(\Delta\varphi_L + \Delta\varphi_{NL}) \qquad (1)$$

$$\Delta\varphi_L = (n_x - n_y)/\beta L \qquad (2)$$

$$\Delta\varphi_{NL} = -\left(\tfrac{1}{3}\right)\gamma PL \cos \alpha_2 \qquad (3)$$

where $\beta=2\pi/L$ is the propagation constant; $\Delta\varphi_L$, $\Delta\varphi_{NL}$ are the linear and nonlinear phase changes; and L, n_x, n_y, γ are the length, linear birefringence coefficient of fast axis and slow axis and nonlinear coefficient of the PCF. P is the instantaneous power of input signal. The quantities α_1 and α_2 which determine the transmission through the NPR structure can be adjusted by changing the two polarization controllers PC1 and PC2.

As shown in Eq. (1), we can plot a cosine curve relation between the transmittivity and the instantaneous power. The value of α_1 and α_2 will not only decide the offset and the amplitude of the cosine curve, but also affect the period of the cosine curve. Tuning PC1 and PC2 properly to change the valuse of α_1 and α_2, then the transmission curve can be varied as shown in Eq. (1). When the rational harmonic mode-locked 30 GHz pulses go through the NPR mechanism, they experience pulse shaping. Therefore, different transmittive curves will result in different pulse shaping mechanisms as suggested by Fig. 3. Fang *et al.* utilized this pulse shaping mechanism induced by the NPR transmission properties and demonstrated flat-top pulse generation in a fiber ring laser [23]. The three transmission curves in Fig. 3 correspond to three differetn α_1 and α_2 sets. If the NPR induced transmittive curve is the dotted line as shown in Fig.3, the low intensity part of the pulse (pulse wings) will have high transmission while the high intensity part (pulse center) will experience low transmission. In this case, the NPR somehow acts as a pulse equalizer which can reduce the intensity fluctuations of the pulse train; however due to the high transmission of pulse wings, it will also broaden the pulse width meanwhile. The dashed line in Fig. 3 corresponds to a state that the high intensity part (pulse center) of the input pulse experiences little loss, which the low intensity parts (pulse wings) undergo high loss. The NPR in this state has the same funtionality as a satruable absorber, thus only leading to pulse compression. Compared to a saturable absorber, the NPR technique has

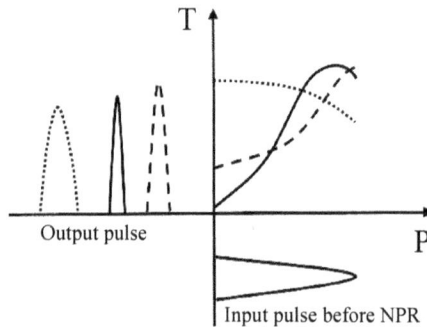

Fig. 3. Conceptual illustration of effects of different transmission curves induced by NPR technique on pulse shaping.

more flexibility in acquiring various transmittive curves. When the tranmission curve is adjusted to the solid line shape, where the pulse wings still undergo great loss, the pulse center will experience a bit higher loss than its adjacent part. Therefore, at this sate, the pulse wdith is narrowed due to the high absorbtion of the leading and trailing edges and if the pulse peak intensity suddenly rises of falls due to the environmental distrubance or mode competition, the transmission loss of the NPR will also rise or fall correspondingly. Thus one can tune the polarization controllers to get different results.

3. Experiment Set-up

Figure 4 shows the experimental setup of our proposed hybrid mode-locked fiber ring laser system. The gain of the fiber laser is provided by an EDFA, which consists of a 23-m long Erbium-doped fiber (EDF) and a 980-nm pump laser diode. A LiNbO$_3$ Mach-Zehnder modulator (MZM) driven by ~10 GHz radio frequency (RF) signal is utilized for rational harmonic mode-locking. Because the loss of the LiNbO$_3$ modulator is polarization sensitive, a polarization controller PC3 is inserted at the input port of the modulator. An optical isolator in the cavity is to ensure unidirectional propagation of the laser mode. A highly nonlinear photonic crystal fiber (PCF), a inline polarizer and two PCs (PC1 and PC2) are used to generate nonlinear polarization rotation effect. If we remove polarization controllers PC1, PC2 and the polarizer, then there is no NPR effect in the cavity. The laser output is coupled out using a 90:10 coupler. All the components are connected by standard single-mode fibers (SSMFs).

This is an all-fiber set up and is easy to construct. The 30 GHz pulse train is generated by rational harmonic mode locking from the MZM driven by the RF signal. With fine tuning of the modulation frequency and the tree polarization controllers, the final results are reached.

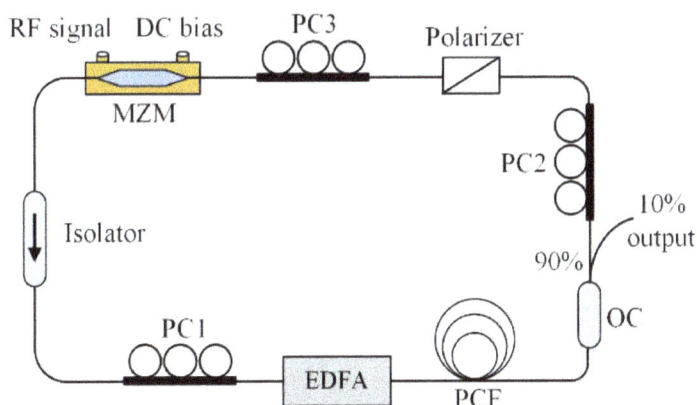

Fig. 4. Experiment setup of the hybrid mode-locked fiber ring laser based on the combination of rational harmonic mode locking and the passive nonlinear polarization rotation technique. EDFA: Er-doped fiber amplifier, MZM: Mach-Zehnder modulator, PC: polarization controller, OC: optical coupler.

4. Numerical Simulation

We have conducted numerical simulation of the hybrid mode-locked fiber laser in MATLAB. For optical pulses with short temporal width, the generalized nonlinear Schrödinger equation (GNLSE) governs their propagation along the fibers [24-26]:

$$\frac{\partial A(z,\tau)}{\partial z} + \frac{\alpha}{2}A(z,\tau) + \sum_{k\geq2}\frac{i^{k-1}}{k!}\beta_k\frac{\partial^k A(z,\tau)}{\partial \tau^k}$$

$$= \frac{g}{2}A(z,\tau) + \frac{g}{2\Omega_g^2}\frac{\partial^2 A(z,\tau)}{\partial \tau^2} + i\gamma A(z,\tau)|A(z,\tau)|^2 \qquad (4)$$

where $A(z,\tau)$ is the slow varying pulse envelope; z is the propagation distance; τ is the time delay parameter; α is the fiber loss and γ is the nonlinear parameter. β_k represent the dispersion coefficients and we include up to the third dispersion parameter in the simulation. Ω_g denotes the gain bandwidth of the EDF and the saturation effect of the EDF is taken into account by expressing the gain factor as $g=g_0/(1+E/E_{sat})$, where g_0 is the small signal gain, E is the pulse energy and E_{sat} is the EDF gain saturation energy. The GNLSE is solved by applying the split-step Fourier method [25-27]. The values for β_2, β_3, γ, E_{sat} and g_0 for EDF used in this numerical simulation are: -0.13×10^{-3} ps^2/m, 0.135×10^{-3} ps^3/m, 3.69W^{-1}km^{-1}, 0.1 pJ and 1.09dB/m. For pulse evolution in the PCF and the SSMF which composes the ring laser cavity, we just simply set $g=0$. Equation 1 models the effect of NPR. The quantities α_1 and α_2 which determine the transmission through the NPR structure can be adjusted by changing the polarization controllers PC1 and PC2. During each round trip after the polarizer, the output pulse envelope $A(z,\tau)_{out} = \sqrt{T(\alpha_1,\alpha_2)}A(z,\tau)_{in}$, in which $T(\alpha_1,\alpha_2)$ as a function of α_1 and α_2 is evaluated by Equation 1-3. In the simulation, we set $\alpha_1=45°$ and explore the effects of various α_2 values. The output coupler is modeled by $A(z,\tau)=RA(z,\tau)$. Since we use a 90/10 coupler in our fiber ring laser system as in Fig. 1, R is 0.9.

The Lithium Niobate Mach-Zehnder modulator driven by an RF signal can be described by the following single-pass transmission function [28-30]:

$$T = \cos^2\left(\pi\frac{V_b+V_m\sin(2\pi f_m\tau)}{V_\pi}\right) \qquad (5)$$

in which V_π is the voltage required for a phase shift of π between the two arms and it's 6 V; V_b is the DC voltage bias; V_m and f_m are the amplitude and frequency of the RF signal, respectively. Simulation is initiated by launching into the system a seed pulse with small amplitude. The pulse evolution within the ring cavity is then iteratively modeled until a steady state is reached after many roundtrips. The asymptotic state is independent on the seed pulse.

Figure 5 depicts the simulated pulse development of the output pulses with only rational harmonic mode-locking and also with the PCF, respectively. As we can see in the figure, after several round trips, the pulse builds up energy and the pulse width becomes stable. The output pulse width is much narrower in the case with the PCF. Figure 6 shows the evolution of pulse width (as it makes the round trips in the cavity) with the PCF scheme as the compressor. The NPR works strongly as a pulse compressor and the generated pulse width is compressed to ~1.88 ps.

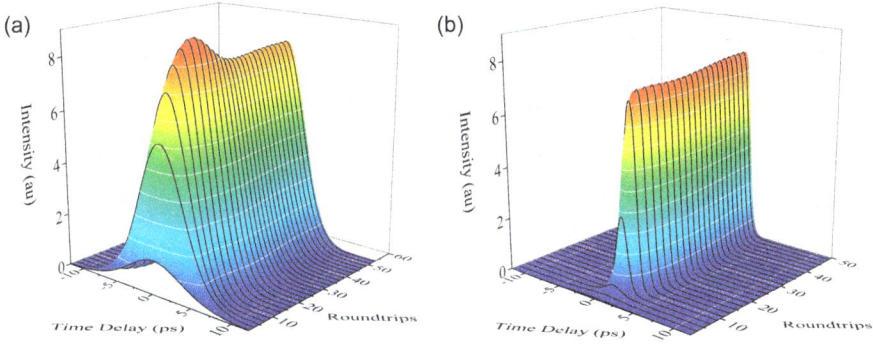

Fig. 5. Simulated results of pulse evolution in the fiber ring laser (a) only with the rational active harmonic mode-locking and (b) implementing the PCF to induce NPR in the cavity, $\alpha_1=45°$ and $\alpha_2=30°$.

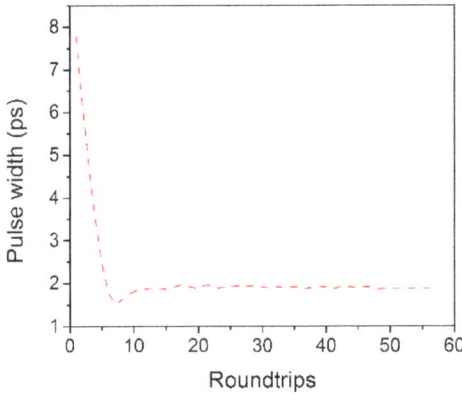

Fig. 6. Numerical simulation results of the evolution of the pulse width in the fiber ring laser with the implementation of the PCF, the α_2 used in the simulation is 30°.

5. Results and Discussions

After carefully tuning the PCs and the frequency of the RF signal, a 30 GHz pulse train with ultrashort pulse width can be generated. When the fundamental frequency of the cavity f_c and the modulation frequency f_m satisfies the condition $f_m=(n+1/p) f_c$, where n and p are both integers, the laser resonator operates in the rational harmonic mode-locking (RHML) regime and the pulse train with a repetition rate of pf_m can be produced. Here in our case, f_m is set at ~10 GHz to realize 3rd order rational harmonics. If we remove polarization controllers PC1, PC2 and the polarizer, we can assume that the NPR has no effect on pulse shaping or there is no NPR in the cavity. Figure 7 shows the auto-correlation trace of the output pulse train with only RHML (a) and with PCF induced NPR (b). Without NPR, the generated pulse train has a calibrated pulse width ~5.8 ps which is very close to our numerical calculated value 5.63 ps. However, with NPR in cavity and after the careful tuning of the PCs, the pulse width of the generated 30 GHz pulse train is shortened to ~1.9 ps. The compressing ratio is as high as 67%. The NPR inside the cavity can greatly improve the pulse shortening mechanism of the fiber ring laser due to the high loss it induced to the pulse wings. The central part of the pulse experiences relatively low loss compared to the pulse wings. This agrees well with the numerical simulation as it shows that with $\alpha_2=30°$, the pulse width could be compressed to 1.88 ps. In Fig. 7, the temporal spacing between the peaks is ~33.2 ps which also confirms the repetition rate at 30 GHz.

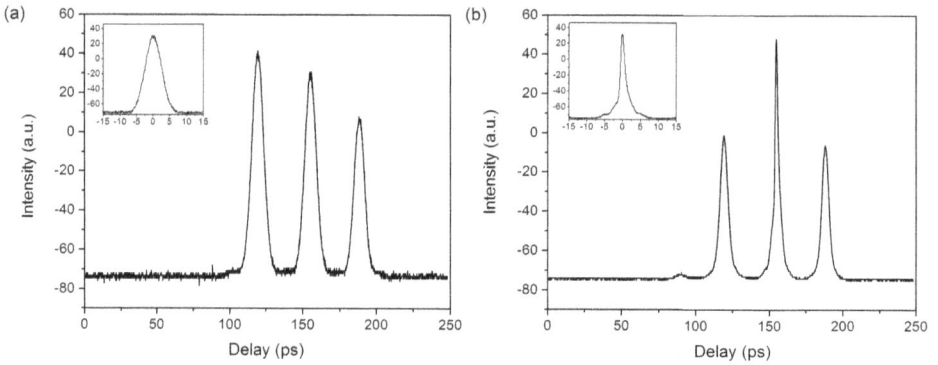

Fig. 7. Auto-correlation trace of the pulse train output from the fiber ring laser. (a) With only rational harmonic mode-locking, the 30 GHz pulse width is ~5.8 ps. (b) With the PCF in the cavity, the 30 GHz pulse width is compressed to ~1.9 ps.

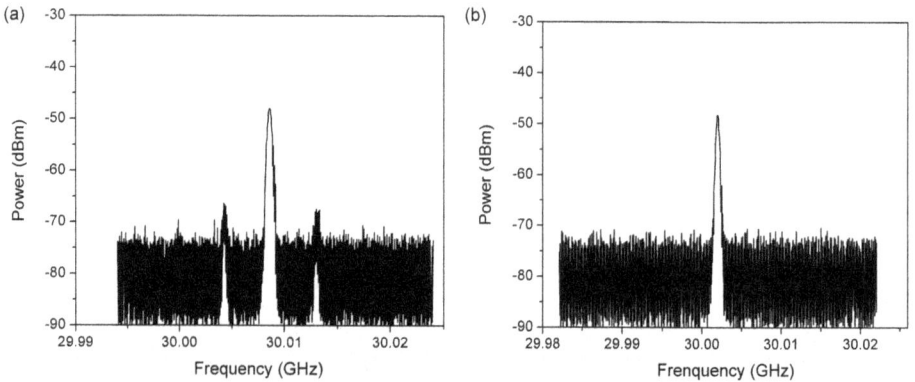

Fig. 8. RF spectra of the generated pulse trains at 30 of the two cases.

Figure 8 compares the RF spectra of the generated 30 GHz pulse train with only RHML (a) and of the PCF induced NPR (b). The highest peak in both figures correspond to the operating frequency of the fiber ring laser which is ~30 GHz. As we can see in Fig. 8(a), without NPR, there are sidebands in the RF spectrum. These are the supermodes which cannot be removed by tuning the modulation frequency f_m and the polarization controller PC3 in the cavity. However, no supermodes are observed in Fig. 8(b) with the presence of NPR in the cavity. The signal-to-noise ratio is over 25 dB after the NPR incorporated. Because the supermodes competition directly leads to pulse fluctuation between adjacent peaks, the reduction of the supermodes in the RF spectrum can improve the stability of optical pulse generation in high order rational harmonic mode locking [6]. The reduced amplitudes of the supermodes in the fiber ring laser is the direct result of the specific transmission property induced by the NPR effect.

6. Conclusion

In conclusion, we have designed and experimentally demonstrated a mode-locked fiber ring laser, which combines rational harmonic mode locking technique and a nonlinear polarization rotation scheme based on a PCF, to generate an optical pulse train with improved stability and shortened pulse width. The laser is operated at 30 GHz using an intra-cavity LiNbO$_3$ Mach-Zehnder modulator at the frequency of ~10 GHz to achieve 3rd rational harmonics. The generated pulse width is compressed from ~5.8 ps to ~1.9 ps and the supermodes noise are reduced for this hybrid scheme. Pulse evolution in the laser ring cavity has also been numerically simulated by solving the generalized nonlinear Schrödinger equation, which shows good agreements with the experimental results.

References

1. M. Nakazawa, T. Hirooka, P. Ruan, and P. Guan, "Ultrahigh-speed "orthogonal" TDM transmission with an optical Nyquist pulse train," Optics express **20**, 1129-1140 (2012).
2. X. Zhang, W. Li, H. Hu, and N. K. Dutta, "High-Speed All-Optical Encryption and Decryption Based on Two-Photon Absorption in Semiconductor Optical Amplifiers," Journal of Optical Communications and Networking **7**, 276-285 (2015).
3. P. V. Mamyshev, S. V. Chernikov, and E. Dianov, "Generation of fundamental soliton trains for high-bit-rate optical fiber communication lines," Quantum Electronics, IEEE Journal of **27**, 2347-2355 (1991).
4. S. Nolte, S. Döring, A. Ancona, J. Limpert, and A. Tünnermann, "High repetition rate ultrashort pulse micromachining with fiber lasers," in *Fiber Laser Applications*, (Optical Society of America, 2011), FThC1.
5. S. Ma, W. Li, H. Hu, and N. K. Dutta, "High speed ultra short pulse fiber ring laser using photonic crystal fiber nonlinear optical loop mirror," Optics Communications **285**, 2832-2835 (2012).
6. C. Wu and N. K. Dutta, "High-repetition-rate optical pulse generation using a rational harmonic mode-locked fiber laser," Quantum Electronics, IEEE Journal of **36**, 145-150 (2000).
7. A. O. Wiberg, C.-S. Brès, B. P. Kuo, J. X. Zhao, N. Alic, and S. Radic, "Pedestal-free pulse source for high data rate optical time-division multiplexing based on fiber-optical parametric processes," Quantum Electronics, IEEE Journal of **45**, 1325-1330 (2009).

8. X. Zhang, H. Hu, W. Li, and N. K. Dutta, "High-repetition-rate ultrashort pulsed fiber ring laser using hybrid mode locking," Applied optics **55**, 7885-7891 (2016).

9. D. Leaird, S. Shen, A. Weiner, A. Sugita, S. Kamei, M. Ishii, and K. Okamoto, "Generation of high-repetition-rate WDM pulse trains from an arrayed-waveguide grating," IEEE Photonics Technology Letters **13**, 221-223 (2001).

10. Y. Fukuchi and J. Maeda, "Characteristics of rational harmonic mode-locked short-cavity fiber ring laser using a bismuth-oxide-based erbium-doped fiber and a bismuth-oxide-based highly nonlinear fiber," Optics express **19**, 22502-22509 (2011).

11. Z. Li, C. Lou, K. T. Chan, Y. Li, and Y. Gao, "Theoretical and experimental study of pulse-amplitude-equalization in a rational harmonic mode-locked fiber ring laser," IEEE Journal of Quantum Electronics **37**, 33-37 (2001).

12. O. Okhotnikov, A. Grudinin, and M. Pessa, "Ultra-fast fibre laser systems based on SESAM technology: new horizons and applications," New journal of physics **6**, 177 (2004).

13. X. He, Z.-b. Liu, and D. Wang, "Wavelength-tunable, passively mode-locked fiber laser based on graphene and chirped fiber Bragg grating," Optics letters **37**, 2394-2396 (2012).

14. H. Hu, X. Zhang, W. Li, and N. K. Dutta, "Hybrid mode-locked fiber ring laser using graphene and charcoal nanoparticles as saturable absorbers," in *SPIE Defense+ Security*, (International Society for Optics and Photonics, 2016), 983630-983630-983638.

15. J. L. Luo, L. Li, Y. Q. Ge, X. X. Jin, D. Y. Tang, S. M. Zhang, and L. M. Zhao, "L-Band Femtosecond Fiber Laser Mode Locked by Nonlinear Polarization Rotation," Photonics Technology Letters, IEEE **26**, 2438-2441 (2014).

16. A. Komarov, H. Leblond, and F. Sanchez, "Passive harmonic mode-locking in a fiber laser with nonlinear polarization rotation," Optics communications **267**, 162-169 (2006).

17. X. Liu, T. Wang, C. Shu, L. Wang, A. Lin, K. Lu, T. Zhang, and W. Zhao, "Passively harmonic mode-locked erbium-doped fiber soliton laser with a nonlinear polarization rotation," Laser physics **18**, 1357-1361 (2008).

18. A. F. Runge, C. Aguergaray, R. Provo, M. Erkintalo, and N. G. Broderick, "All-normal dispersion fiber lasers mode-locked with a nonlinear amplifying loop mirror," Optical Fiber Technology **20**, 657-665 (2014).

19. W. Li, H. Hu, X. Zhang, S. Zhao, K. Fu, and N. K. Dutta, "High-speed ultrashort pulse fiber ring laser using charcoal nanoparticles," Applied Optics **55**, 2149-2154 (2016).

20. X. Zhang, H. Hu, W. Li, and N. K. Dutta, "Supercontinuum generation in dispersion-varying microstructured optical fibers," in *SPIE Defense+ Security*, (International Society for Optics and Photonics, 2016), 98340F-98340F-98347.

21. J. M. Dudley and J. R. Taylor, "Ten years of nonlinear optics in photonic crystal fibre," Nature Photonics **3**, 85-90 (2009).

22. P. S. J. Russell, "Photonic-crystal fibers," Journal of lightwave technology **24**, 4729-4749 (2006).

23. X. Fang, P. Wai, C. Lu, and J. Chen, "Flattop pulse generation based on the combined action of active mode locking and nonlinear polarization rotation," Applied optics **53**, 902-906 (2014).

24. J. Jeon, J. Lee, and J. H. Lee, "Numerical study on the minimum modulation depth of a saturable absorber for stable fiber laser mode locking," JOSA B **32**, 31-37 (2015).

25. G. P. Agrawal, *Nonlinear fiber optics* (Academic press, 2007).

26. X. Zhang, H. Hu, W. Li, and N. K. Dutta, "Mid-infrared supercontinuum generation in tapered As2S3 chalcogenide planar waveguide," Journal of Modern Optics **63**, 1965-1971 (2016).

27. H. Hu, X. Zhang, W. Li, and N. K. Dutta, "Simulation of octave spanning mid-infrared supercontinuum generation in dispersion-varying planar waveguides," Applied Optics **54**, 3448-3454 (2015).

28. H. Dong, H. Sun, G. Zhu, Q. Wang, and N. Dutta, "Clock recovery using cascaded LiNbO3 modulator," Optics express **12**, 4751-4757 (2004).

29. W. Li, H. Hu, X. Zhang, and N. K. Dutta, "High Speed All Optical Logic Gates Using Binary Phase Shift Keyed Signal Based On QD-SOA," International Journal of High Speed Electronics and Systems **24**, 1550005 (2015).

30. X. Zhang and N. K. Dutta, "Effects of two-photon absorption on all optical logic operation based on quantum-dot semiconductor optical amplifiers," Journal of Modern Optics **65**, 166-173 (2018).

Chirped Superlattices as Adjustable Strain Platforms for Metamorphic Semiconductor Devices

Md Tanvirul Islam

Electrical and Computer Engineering Department, 371 Fairfield Way,
Unit 4157, Storrs, CT 06269-4157, USA
md.t.islam@uconn.edu

Xinkang Chen

Electrical and Computer Engineering Department, 371 Fairfield Way,
Unit 4157, Storrs, CT 06269-4157, USA
xinkang.chen@uconn.edu

Tedi Kujofsa

Electrical and Computer Engineering Department, 371 Fairfield Way,
Unit 4157, Storrs, CT 06269-4157
tedi.kujofsa@gmail.com

John E. Ayers[*]

Electrical and Computer Engineering Department, 371 Fairfield Way,
Unit 4157, Storrs, CT 06269-4157, USA
john.ayers@uconn.edu

Chirped superlattices are of interest as buffer layers in metamorphic semiconductor device structures, because they can combine the mismatch accommodating properties of compositionally-graded layers with the dislocation filtering properties of superlattices. Important practical aspects of the chirped superlattice as a buffer layer are the surface strain and surface in-plane lattice constant. In this work two basic types of InGaAs/GaAs chirped superlattice buffers have been studied. In design I (composition modulated), the average composition is varied by modulating the composition of one of the two layers in the superlattice period, but the individual layer thicknesses were fixed. In design II (thickness modulated), the individual layer thicknesses were modulated, but the compositions were fixed. In this paper the surface strain and surface in-plane lattice constant for these chirped superlattices are presented as functions of the top composition and period for each of these basic designs.

Keywords: Electrical circuit model; strain; step-grading; InGaAs/GaAs; chirped superlattices; compositionally-graded.

[*]Corresponding author.

1. Introduction

Metamorphic realization of mismatched semiconductor devices such as high electron mobility transistors (HEMTs), light-emitting diodes (LEDs), and multi-junction solar cells is of great interest due to the flexibility it affords in designing layer compositions and thicknesses. A drawback of metamorphic growth is the inherent incorporation of misfit dislocations and their associated threading defects, which can degrade device performance. It is possible to minimize these effects by using a suitable buffer layer grown between the device and its mismatched substrate, and the linearly-graded buffer has been the most-studied approach. Use of a grading coefficient less than 500 cm^{-1} (where the average grading coefficient is defined as the change in fractional mismatch divided by the thickness) has been found to greatly reduce the threading dislocation density at the top of such a linearly-graded buffer [1]. On the other hand, superlattices have been used either alone or in conjunction with graded layers to filter threading dislocations [1]. In this work, it is proposed that chirped superlattices, in which the average composition is graded with distance, could embody the benefits of both linearly-graded and superlattice buffers. When applying such a buffer for a device structure, it is desirable to be able to control the in-plane lattice constant at the top of the buffer, so it can match the intended device layer and thereby avoid the introduction of new dislocations at the interface between the buffer and the device. Two basic types of InGaAs/GaAs chirped superlattice buffers have been considered. In design I (composition modulated), the average composition was varied by controlling the composition of one of the two layers in the superlattice period, but the individual layer thicknesses were fixed as shown in Fig. 1(a). This design involved five periods of dimension λ. Each period included two In$_x$Ga$_{1-x}$As layers of thickness $\lambda/2$; the first layer in each period had an indium composition of $x_A = x_{top}$ while the second layer in each period had a composition which varied linearly with the number of the period according to $x_B = x_{top}(N-1)/4$, where N is the number of the period. In design II

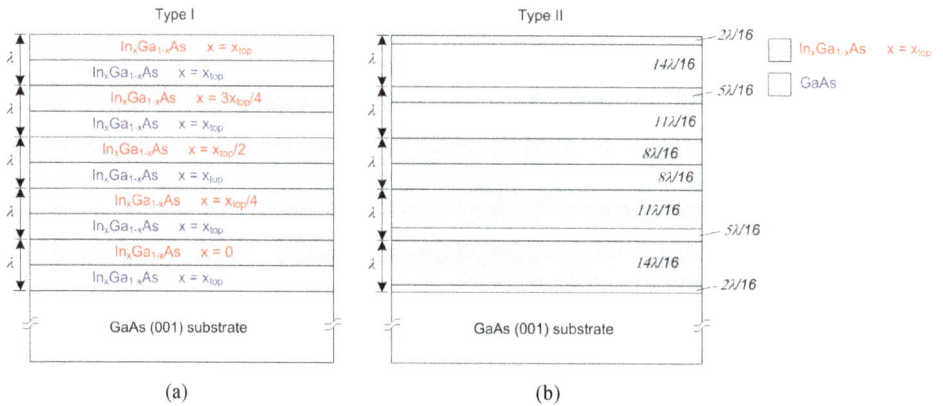

Fig. 1. Chirped superlattice structures considered in this study. The type I structure (a) utilizes composition modulation while the type II structure (b) utilizes thickness modulation.

(thickness modulated), the individual layer thicknesses were varied, but the compositions were fixed as shown in Fig. 1(b). This design also involved five periods of dimension λ. Within each period the first layer was GaAs while the second layer was $In_xGa_{1-x}As$ with an indium composition of x_{top}; however, the thicknesses of the individual layers were varied according to $h_A = \lambda(3N-1)/16$ for the GaAs layer and $h_B = \lambda(17-3N)/16$ for the $In_xGa_{1-x}As$ layer, where N is the number of the period. In the following study the parameters λ and x_{top} have been varied in each type of chirped superlattice, and the resulting equilibrium surface strain and surface in-plane lattice constant are reported.

2. Theory

In this work the equilibrium strain and misfit dislocation densities were determined for composition-modulated and thickness-modulated InGaAs/GaAs (001) chirped super-lattices using the electric circuit model (ECM) for strained-layer epitaxy, as described by Kujofsa and Ayers [2]. The basis of the ECM is energy minimization in a graded or multilayered structure. Any such structure may be approximated by a stack of uniform composition sublayers, and each of these sublayers is represented by an analogous subcircuit comprising a resistor, an independent current source, an independent voltage source, and an ideal diode. For example, Fig. 2 illustrates a three-layer structure and its analogous circuit model. The equilibrium strains in the epitaxial structure have the same numerical values as the node voltages in the analogous circuit. In the absence of coherently-strained interfaces, the equations governing the analogous circuit model and its behavior are the following.

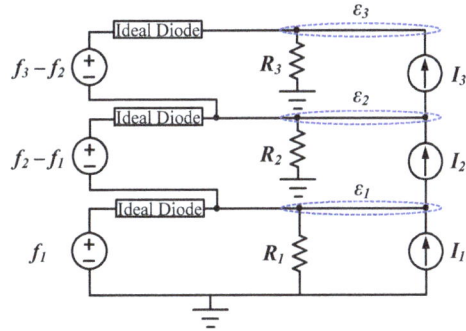

Fig. 2. In-plane strain in a multilayered heterostructure and the corresponding electrical circuit model.

The resistor in the nth subcircuit, R_n, has a numerical value given by

$$R_n \quad \leftrightarrow \quad \frac{1}{2Y_n h_n}, \tag{1}$$

where Y_n is the biaxial modulus and h_n is the thickness for the nth sublayer. The double-sided arrow indicates that " R_n is analogous to $1/(2Y_n h_n)$," and although the two quantities of either side of the double arrow possess different engineering units, they have the same numerical value. The current source I_n in the nth subcircuit has a numerical value given by

$$I_n \quad \leftrightarrow \quad \frac{f_n - f_{n-1}}{|f_n - f_{n-1}|} \frac{G_n b_n \left(1 - v_n \cos^2 \theta\right)}{2\pi\left(1 - v_n\right)\sin \lambda \sin \phi} \left[\ln\left(\sum_{j=n}^{N} \frac{h_j}{b_n}\right) + 1 \right], \tag{2}$$

where f_n is the lattice mismatch, G_n is the shear modulus, b_n is the length of the Burgers vector, v_n is the Poisson ratio, θ is the angle between the Burgers vector and the line vector for dislocations, λ is the angle between the Burgers vector and the direction in the interface which is perpendicular to the intersection of the glide plane and the interface, and ϕ is the angle between the glide plane and interface. Applying the node voltage approach, the node voltage at the nth node of the analogous circuit (analogous to the strain in the nth sublayer) is given by

$$\varepsilon_n \quad \leftrightarrow \quad V_n = \begin{cases} R_n \cdot \left(I_n - I_{n+1}\right), & 1 \le n < N \\ R_n \cdot I_n, & n = N \end{cases} . \tag{3}$$

The primary advantage of using the circuit model analogy is that it leverages the supernode and ideal diode concepts from circuit theory, which do not exist in mechanical circuits, in order to greatly simplify the analysis and enable intuitive design of strained-layer structures. When coherently-strained interfaces are involved, the associated ideal diodes turn on, and the resulting node voltages are determined using the concept of the supernode from circuit theory. The existence of the supernode modifies the Kirchhoff current law equations for the nodes involved, and therefore the resulting voltages. If the supernode is bounded inclusively by sublayers σ and ω, then the equilibrium strain (node voltage) in the bottom layer of the supernode has a numerical value given by

$$\varepsilon_\sigma \quad \leftrightarrow \quad V_\sigma = \left[\left(I_\sigma - I_{\omega+1}\right) - \sum_{j=\sigma}^{\omega} \frac{f_j - f_\sigma}{R_j}\right] R_{SN}, \tag{4}$$

where the equivalent parallel resistance of the supernode (R_{SN}) is defined as the equivalent resistance for a series of resistors in parallel, $R_{SN} = R_\sigma \| ..... \| R_\omega$, and is given by

$$R_{SN} = \left(\sum_{j=\sigma}^{\omega} \frac{1}{R_j}\right)^{-1} . \tag{5}$$

The in-plane strain (analogous to node voltage) at each sublayer corresponding to the supernode is then determined by adding the appropriate sum of independent voltage sources to the voltage at the bottom of the supernode. In other words, the node voltage (or the equivalent in-plane strain) of each sublayer (not including the bottom layer) of the supernode is determined from

$$\varepsilon_i = \varepsilon_\sigma + \sum_{j=\sigma+1}^{i} f_j - f_{j-1} \quad \leftrightarrow \quad V_i = V_\sigma + \sum_{j=\sigma+1}^{i} VS_j, \quad \sigma < i \le \omega, \tag{6}$$

where the voltage source in the subcircuit representing the nth sublayer, VS_n, is analogous to the lattice mismatch difference:

$$VS_n \quad \leftrightarrow \quad f_n - f_{n-1}, \tag{7}$$

and $f_0 \equiv 0$ corresponds to the substrate. Further details of the ECM are contained in reference [2].

The material parameters used in this work for GaAs, InAs, and $In_xGa_{1-x}As$ are summarized in Table 1.

Table 1. Material Properties for GaAs, InAs, and the alloy $In_xGa_{1-x}As$.

Parameter	GaAs	$In_xGa_{1-x}As$	InAs
a (nm)	0.56534	0.56534 + x(0.04050)	0.60584
b (nm)	0.39976	0.39976 + x(0.02864)	0.42839
Y (GPa)	124	124 – x(45)	79
G (GPa)	32	32 – x(13)	19
ν	0.31	0.31 + x(0.04)	0.35
θ (degrees)	60	60	60
λ (degrees)	60	60	60
φ (degrees)	35.26	35.26	35.26

3. Results and Discussion

The equilibrium strain was analyzed in chirped superlattices of the two types shown in Fig. 1. For example, Fig. 3 illustrates the strain profile for chirped superlattices with $\lambda = 40$ nm and $x_{top} = 0.3$ for the case of type I (a) and type II (b). Similar profiles were found for a number of values of λ and x_{top}, and from them, the average strain, surface strain, and surface lattice constant for each type of superlattice were determined and plotted. Figures 4(a) and 4(b) show the average strain as a function of the top composition and period, respectively. The variation with composition is related to changes in the critical layer thickness and elastic constants, while there is the expected inverse relationship between period (and therefore total thickness, with a fixed number of periods) and average strain. The surface strain variation with top composition and period are shown in Figs. 5(a) and 5(b), respectively. These characteristics are very different for type I and type II chirped superlattices, because the type II structure always has a thin GaAs layer on top. Finally, Figs. 6(a) and 6(b) show the surface in-plane lattice constant as a function of top composition and period, respectively, for both types of chirped superlattices. These results show that the in-plane lattice constant can be varied up to 0.584 nm, or up to an indium equivalent of 0.46, if the x_{top} is varied up to 0.5 as considered here. Therefore the chirped superlattices with x_{top} up to 0.5 can serve as platforms for the growth of relaxed InGaAs device layers with compositions up to 0.46 indium.

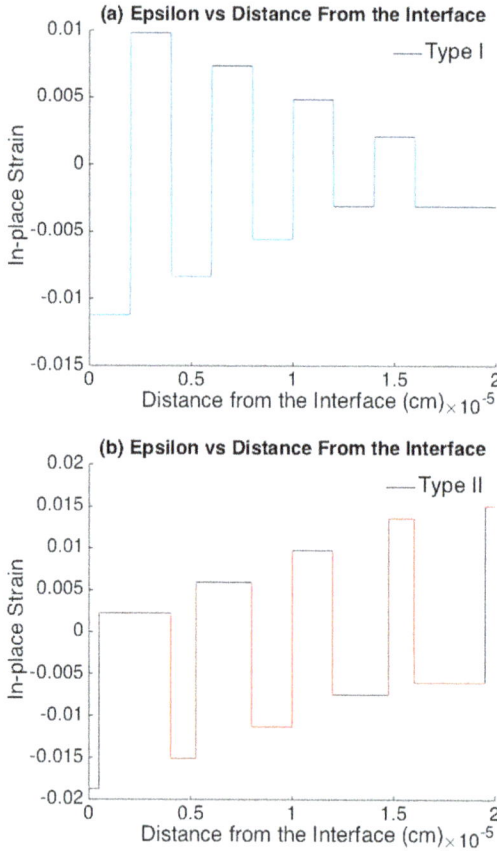

Fig. 3. In-plane strain profiles for representative type I (a) and type II (b) chirped superlattices. The period was fixed at 40 nm while x_{top} was fixed at 0.3.

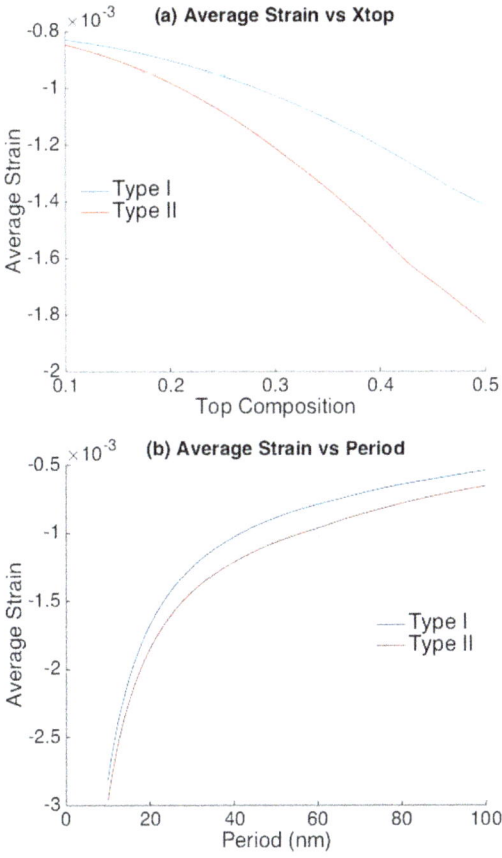

Fig. 4. Average strain for type I and type II chirped superlattices as a function of top composition with the period fixed at 40 nm (a) and as a function of period with the top composition fixed at 0.3 (b).

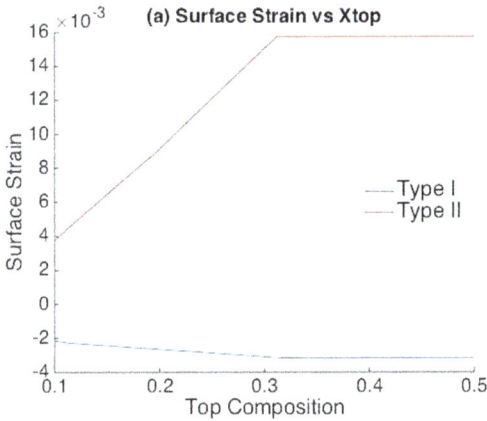

Fig. 5. Surface strain for type I and type II chirped superlattices as a function of top composition with the period fixed at 40 nm (a) and as a function of period with the top composition fixed at 0.3 (b).

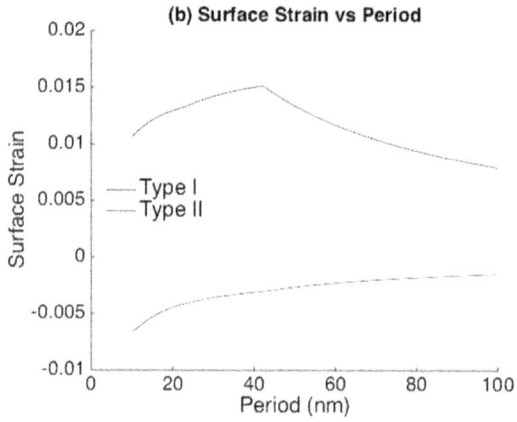

Fig. 5. (*Continued*)

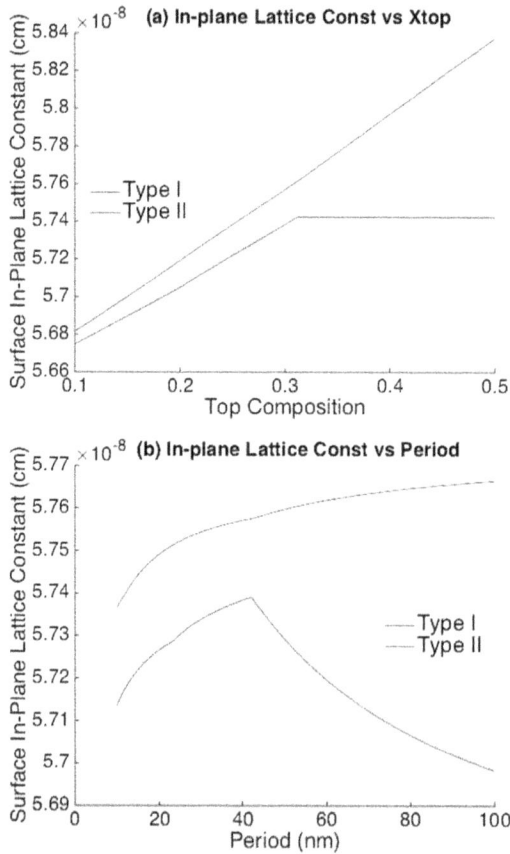

Fig. 6. Surface in-plane lattice constant for type I and type II chirped superlattices as a function of top composition with the period fixed at 40 nm (a) and as a function of period with the top composition fixed at 0.3 (b).

4. Conclusion

We have shown that InGaAs chirped superlattices of type I (composition modulated) or type II (thickness modulated) may be designed to have a wide range of surface in-plane lattice constant up to 0.584 nm if x_{top} is varied up to 0.5, and can therefore serve as growth platforms for relaxed InGaAs having an indium mole fraction up to 0.46.

References

1. T. Kuech, Ed., Handbook of Crystal Growth, Second Edition, Vol. 3: Thin Films and Epitaxy, Ch. 25, "Low temperature and metamorphic buffer layers," by J. E. Ayers (Elsevier, 2014).
2. T. Kujofsa and J. E. Ayers, Semicond. Sci. Technol., 31, 115014 (2016).

Dynamical X-Ray Diffraction Analysis of a GaAs/In₀.₃Ga₀.₇As Single Quantum Well Grown on a GaAs (001) Substrate

F. A. Althowibi*

Electrical Engineering Department, Taif University,
Taif, P.O. Box: 888, 21974, Saudi Arabia
fhdamer@gmail.com

J. E. Ayers

Electrical and Computer Engineering Department, University of Connecticut,
Storrs, Connecticut 06269-4157, USA
john.ayers@uconn.edu

We applied the mosaic crystal model to calculate the dynamical x-ray rocking curves for a coherently-strained GaAs/In₀.₃Ga₀.₇As/GaAs single quantum well grown epitaxially on a GaAs (001) substrate for a number of reflection profiles, including 004, 113, 224, 044 and 444 reflections. We show that it is possible to estimate the threading dislocation density in the quantum well, and therefore detect the pseudomorphic-metamorphic transition, using the widths or normalized intensities of the primary quantum well Bragg peak, or using the widths of the Pendellösung fringes associated with the quantum well structure.

Keywords: X-ray diffraction; dynamical diffraction; metamorphic heterostructures; dislocations; quantum well.

1. Introduction

Structures involving single or multiple quantum wells are of interest for a number of semiconductor device applications [1–5]. Structural analysis of quantum well structures is well established, and enables the depth profiling of strain and composition using high-resolution x-ray rocking curves. Using high-resolution x-ray rocking curves as a non-destructive tool for characterizing threading dislocation density may take the structural analysis of quantum wells a step further. Though the use of quantum-well structures is important for the realization of a number of semiconductor devices, these quantum wells involve lattice-mismatched structures and may introduce threading dislocations. These dislocations are known to degrade device performance, so that characterization of threading dislocations in such structures may be essential for semiconductor device optimization.

---

*Corresponding author.

In the published literature both kinematical [6–10] and dynamical [11–16] approaches have been described for the depth profiling of strain and composition, but these simulations ignore the effects of defects and may not be used to characterize dislocation densities. Recently, the dynamical simulations have been extended to the depth profiling of threading dislocation density, using the phase invariant [17] and mosaic crystal [18] models.

In this study, a study of simulated dynamical x-ray rocking curves was conducted on a coherently-work strained GaAs/$In_{0.3}Ga_{0.7}As$/GaAs single quantum well assumed to be grown epitaxially on a GaAs (001) substrate. The mosaic crystal model was utilized for a number of reflection cases, including the 004, 113, 224, 044 and 444 reflections for the case of Cu $k\alpha_1$ radiation. We show that it is possible to estimate the threading dislocation density in the quantum well by analysis of the width and normalized intensity of the primary quantum well Bragg peak, or by using the widths of the Pendellösung fringes associated with the quantum well structure. This analysis therefore provides an indication of metamorphic growth (the presence of misfit dislocations) which can be more sensitive than the traditional approach based on x-ray measurements of strain.

2. Theory

The dynamical diffraction profile for a perfect, unstrained, and infinitely thick semiconductor crystal can be described by the Takagi-Taupin equation [11–13] as

$$-i\frac{dX}{dT} = X^2 - 2\eta X + 1 \tag{1}$$

where X is the complex scattering amplitude, η is the deviation parameter, and T is the thickness parameter calculated, which for a substrate crystal is given by

$$T = h\frac{\pi\Gamma\sqrt{F_{HS}\,F_{\bar{H}S}}}{\lambda\,\sqrt{|\gamma_0\gamma_H|}} \tag{2}$$

where h is the depth measured from the diffracting surface, and F_{HS} and $F_{\bar{H}S}$ are the substrate structure factors for the hkl and $\bar{h}\bar{k}\bar{l}$ reflections and $\Gamma = r_e\lambda^2/(\pi V)$, where r_e is the classical electron radius, $2.818 \times 10^{-5}\dot{A}$, λ is the x-ray wavelength, V is the unit cell volume, and γ_0 and γ_H are the direction cosines for the incident and reflected beams with respect to the inward surface normal.

The Darwin-Prins formula [19] describes the solution to the Takagi-Taupin equation [11–13], and the resulting substrate surface scattering amplitude is then given by

$$X_0 = \eta_s - Sign(\eta_s)\sqrt{\eta_s^2 - 1} \tag{3}$$

where the deviation parameter for the substrate is given by

$$\eta_s = \frac{-(\gamma_0/\gamma_H)(\theta - \theta_{BS})sin(2\theta_{BS}) - 0.5(1 - \gamma_0/\gamma_H)\Gamma F_{0S}}{\sqrt{|\gamma_0/\gamma_H|}\,C\Gamma\,\sqrt{F_{HS}\,F_{\bar{H}S}}} \tag{4}$$

where θ_{BS} is the Bragg angle for the substrate, θ is the actual angle of incidence on the diffracting planes, F_{0S} is the substrate structure factor for 000 scattering, and C is the polarization factor.

The dynamical diffraction profile for a dislocation-free single layer or device structure has been previously well-described in references [18, 24], while the diffused scattering associated with dislocations is considered here. For a defected semiconductor device heterostructures, the crystal is distorted by the presence of dislocations casing variations in both orientations and interplanar spacings [20–22]. To account for causing such distortion, the structure is considered to be made up of mosaic blocks in which dislocations may cause tilting and Bragg angle variations within the blocks. For the mosaic crystal model calculations used here, the semiconductor structure was modeled as an array of $N_\alpha \times N_\beta$ crystallites. The N_α dimension accounts for the angular mosaicity while the N_β dimension accounts for d-spacing mosaicity. Each crystallite is divided into a number of sublayers within its depth, and the Takagi-Taupin equation is used to solve iteratively for the surface scattering amplitude of each crystallite. For the nth lamina of the ij^{th} crystallite, the scattering amplitudes are calculated iteratively by:

$$X_{ijn} = \eta_{ijn} + \sqrt{\eta_{ijn}^2 - 1} \frac{(S_{ij1n} + S_{ij2n})}{(S_{ij1n} - S_{ij2n})} \tag{5}$$

where

$$S_{ij1n} = \left(X_{ijn-1} - \eta_{ijn} + \sqrt{\eta_{ijn}^2 - 1}\right) exp\left(-iT_n\sqrt{\eta_{ijn}^2 - 1}\right) \tag{6}$$

$$S_{ij2n} = \left(X_{ijn-1} - \eta_{ijn} + \sqrt{\eta_{ijn}^2 - 1}\right) exp\left(iT_n\sqrt{\eta_{ijn}^2 - 1}\right) \tag{7}$$

$$\eta_{ijn} = \frac{-(\gamma_0/\gamma_H)(\theta - \theta_{Bn} + \alpha_i - \beta_j)sin(2\theta_{Bn}) - 0.5(1 - \gamma_0/\gamma_H)\Gamma F_{0n}}{\sqrt{|\gamma_0/\gamma_H|}\, C\Gamma \sqrt{F_{Hn} F_{\bar{H}n}}} \tag{8}$$

and

$$T_n = h_n \frac{\pi\Gamma\sqrt{F_{Hn} F_{\bar{H}n}}}{\lambda \sqrt{|\gamma_0\gamma_H|}} \tag{9}$$

where X_{ijn} and η_{ijn} are the complex scattering amplitude and deviation parameter for the nth sublayer of the ij^{th} crystallite, and T_n is the thickness parameter for the nth sublayer. S_{ij1n} and S_{ij2n} represent the scattering amplitudes at the top and the bottom of the nth sublayer of the ij^{th} crystallite, h_n is the thickness, and F_{0n}, F_{Hn} and $F_{\bar{H}n}$ are the 000, hkl and $\bar{h}\bar{k}\bar{l}$ structure factors for the nth sublayer. Once the surface scattering amplitude, X_{ijn}, is calculated for the top sublayer in each of the $N_\alpha \times N_\beta$ crystallites, the diffracted intensity for the mosaic crystal is found from a weighted sum of the individual intensities,

$$I = \sum_i \sum_j |X_{ijn}|^2 \cdot W_{\alpha i} \cdot W_{\beta j}, \tag{10}$$

where $W_{\alpha i}$ and $W_{\beta j}$ are the weighting functions accounting for tilting and Bragg angle variations, and their distributions are assumed to be Gaussian given by [21-24]

$$W_{\alpha i} = exp(-\alpha_i^2/2\sigma_\alpha^2) \tag{11}$$

and

$$W_{\beta j} = exp(-\beta_j^2/2\sigma_\beta^2) \tag{12}$$

where $\alpha_i = -N_\sigma \sigma_\alpha + iN_\alpha \sigma_\alpha/2N_\sigma$ and $\beta_j = -N_\sigma \sigma_\beta + jN_\beta \sigma_\beta/2N_\sigma$, where N_σ is the number of standard deviations used in the two distributions and i and j are integers. The standard deviations are [21-22] given by

$$\sigma_\alpha = b \sqrt{\pi \, (TDD)/2} \tag{13}$$

and

$$\sigma_\varepsilon = 0.127b \cdot \sqrt{(TDD)\left|ln\left(2 \times 10^{-7} cm\sqrt{TDD}\right)\right|} \cdot tan(\theta_B) \tag{14}$$

where TDD is the threading dislocation density and b is the Burgers vector for the dislocations. Such an approach provides an x-ray characterization of metamorphic structures, allowing for depth profiling of strain, composition, and dislocation density.

3. Results and Discussion

Dynamical diffraction profiles were calculated for a GaAs / In$_{0.3}$Ga$_{0.7}$As / GaAs single quantum-well on a GaAs (001) substrate for the cases with and without dislocations, using the mosaic crystal model for a number of reflection profiles including the 004, 113, 224, 044, and 444 for the case of Cu kα_1 radiation ($\lambda = 0.1540594$ nm). Here, coherent strain is

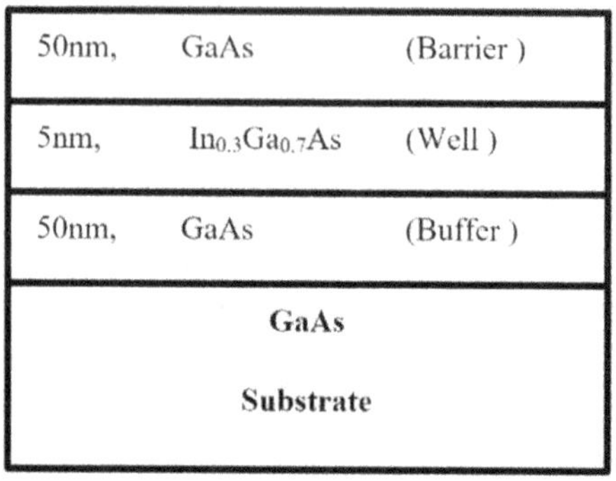

50nm,	GaAs	(Barrier)
5nm,	In$_{0.3}$Ga$_{0.7}$As	(Well)
50nm,	GaAs	(Buffer)
	GaAs	
	Substrate	

Fig. 1. Schematic structure for 50 nm GaAs / 5 nm In$_{0.3}$Ga$_{0.7}$As / 50 nm GaAs single quantum well grown on a GaAs (001) substrate.

assumed and all material parameters for GaAs, InAs, and InGaAs and their associated structure factors were obtained from the references [25, 26]. Figure 1 illustrates a 50 nm GaAs / 5 nm In$_{0.3}$Ga$_{0.7}$As / 50 nm GaAs single quantum-well structure on a GaAs (001) substrate considered for this study.

First, the dynamical diffraction analysis was conducted assuming the dislocation-free case. Figure 2 shows the 004 high-resolution diffraction profile from a pseudomorphic 50 nm GaAs / 5 nm In$_{0.3}$Ga$_{0.7}$As / 50 nm GaAs single quantum-well structure on a GaAs (001) substrate. This profile shows a strong diffraction peak corresponding to the substrate as well as Pendellösung fringes corresponding to the single quantum well structure. Several of the Pendellösung fringes were selected for further analysis as shown in Fig. 2. It should be noted that this choice is not unique, and the analysis of other fringes would be qualitatively similar.

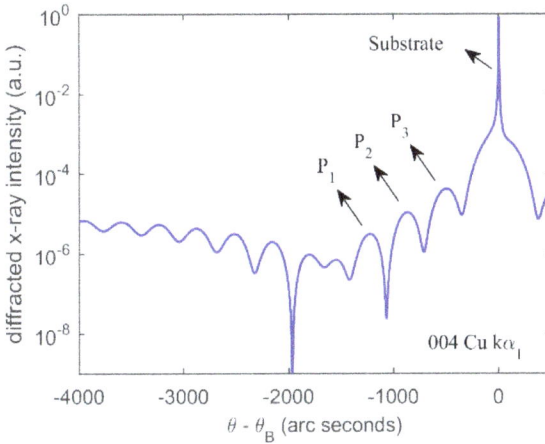

Fig. 2. Calculated 004 rocking curve of a pseudomorphic 50 nm GaAs / 5 nm In$_{0.3}$Ga$_{0.7}$As / 50 nm GaAs single quantum-well structure on a GaAs (001) substrate.

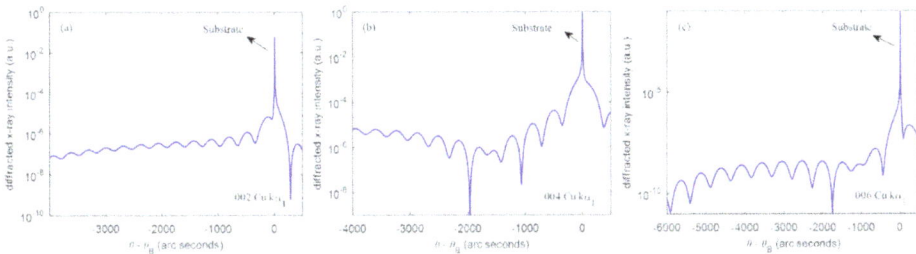

Fig. 3. Calculated rocking curves for a pseudomorphic 50 nm GaAs / 5 nm In$_{0.3}$Ga$_{0.7}$As / 50 nm GaAs single quantum-well on a GaAs (001) substrate for different reflection orders: (a) 002; (b) 004; and (c) 006.

Figure 3 shows the calculated 002, 004, and 006 rocking curves, respectively, for the same pseudomorphic single quantum-well structure detailed in Fig. 1. Here, the diffraction profiles for all three cases are similar, but the 004 reflection yields the highest peak intensity; this is because the 002 and 006 cases both exhibit lower structure factors. Therefore, the 004 reflection profile is the best candidate among symmetrical diffraction cases to analyze the threading dislocation density.

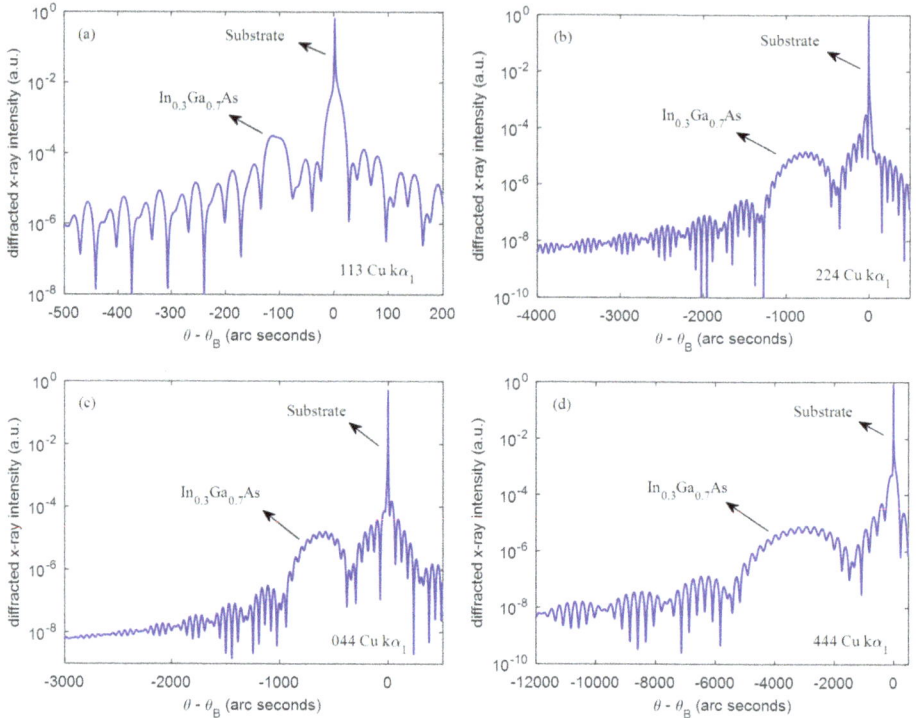

Fig. 4. Calculated rocking curves for a pseudomorphic 50 nm GaAs / 5 nm $In_{0.3}Ga_{0.7}As$ / 50 nm GaAs single quantum well on a GaAs (001) substrate for different reflection profiles: (a) 113; (b) 224; (c) 044; and (d) 444.

To determine the behavior of asymmetrical reflections, Fig. 4 shows the calculated 113, 224, 044 and 444 rocking curves, respectively, from a pseudomorphic 50 nm GaAs / 5 nm $In_{0.3}Ga_{0.7}As$ / 50 nm GaAs single quantum-well grown on GaAs (001) substrate. Interestingly, the results showed an additional diffraction peak corresponding to the single quantum-well layer, strongly observed in the case of 113 (Fig. 4(a)) and weakly observed in the large Bragg angle with broadened diffraction peak as seen in the case of 444 reflection (Fig. 4(d)).

Next the effect of threading dislocations was considered. Figure 5 depicts the calculated 004 rocking curve from a coherently-strained 50 nm GaAs / 5 nm $In_{0.3}Ga_{0.7}As$ / 50 nm GaAs single quantum-well grown on GaAs (001) substrate, with various uniformly assumed threading dislocations: 0, 10^5, 10^6, 10^7, 10^8, and 10^9 cm^{-2}. As the dislocation density is increased, the individual fringes lose intensity and broaden. Either of these two effects may be used to estimate the threading dislocation density.

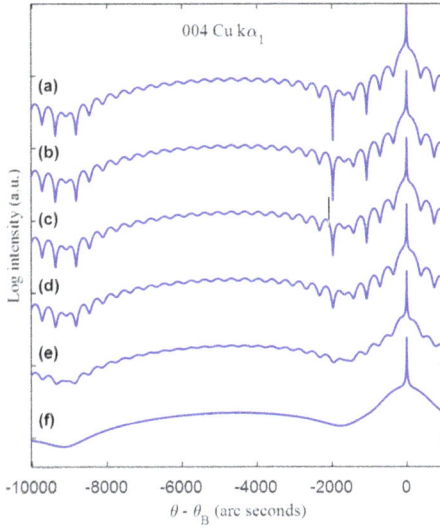

Fig. 5. Calculated 004 rocking curves for a coherently-strained 50 nm GaAs / 5 nm In$_{0.3}$Ga$_{0.7}$As / 50 nm GaAs single quantum well on a GaAs (001) substrate with different uniform dislocation densities: (a) 0 cm^{-2}; (b) 10^5 cm^{-2}; (c) 10^6 cm^{-2}; (d)10^7 cm^{-2}; (e) 10^8 cm^{-2}; and (f)10^9 cm^{-2}.

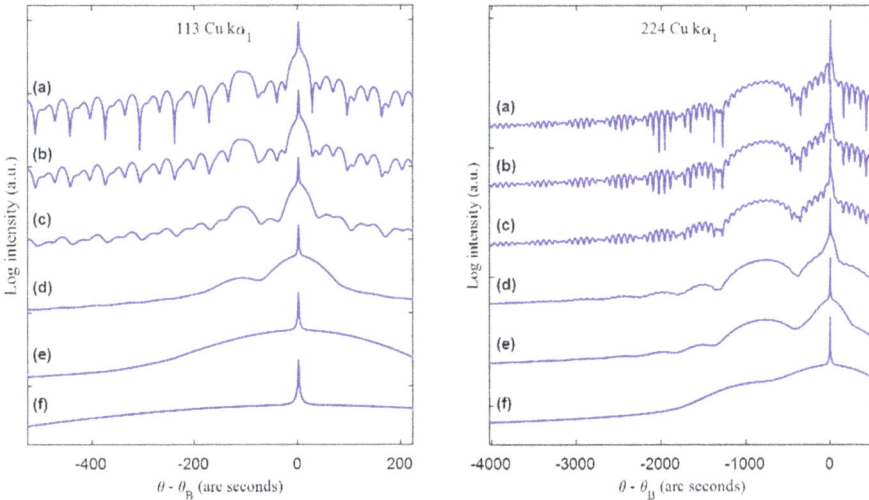

Fig. 6. Calculated 113 and 224 rocking curve for a coherently-strained 50 nm GaAs / 5 nm In$_{0.3}$Ga$_{0.7}$As / 50 nm GaAs single quantum well on a GaAs (001) substrate with different uniform dislocation densities: (a) 0 cm^{-2}; (b) 10^5 cm^{-2}; (c) 10^6 cm^{-2}; (d)10^7 cm^{-2}; (e) 10^8 cm^{-2}; and (f)10^9 cm^{-2}.

Because rocking curve sensitivity toward the thread density should vary with the Bragg angle, we investigated various hkl reflections. Figure 6 depicts the calculated 113 and 224 rocking curves from the same coherently-strained single quantum-well structure, and subjected to threading dislocation variations. It can be seen that the primary Bragg peak for the quantum well and the Pendellösung fringes broaden monotonically with increasing threading dislocation density. Therefore, by correlating the widths of these features with the dislocation density, it should be possible to characterize the latter experimentally.

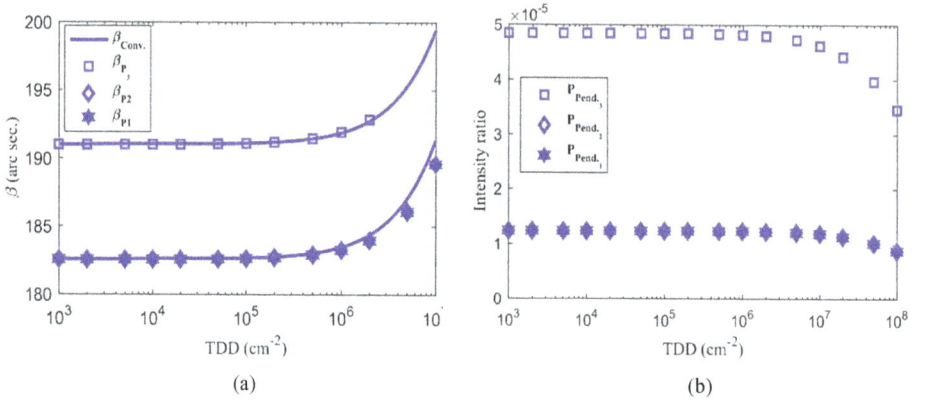

(a)   (b)

Fig. 7. Calculated 004 rocking curve widths (a) and intensity ratios (b) for the selected Pendellösung fringes (as indicated in Fig. 2), as a function of the threading dislocation density (TDD). Here, P_3 is near the substrate diffraction peak (β_{P3} open-square) followed by P_2 (β_{P2}, open-diamond), and P_1 (β_{P1}, star), and all widths in (a) are compared with the width predicted by convolution mode (solid).

Both the widths and normalized intensities of Pendellösung fringes were studied as functions of the threading dislocation density. Figure 7 shows the widths (a) and normalized intensities (b) for the selected fringes determined from Fig. 2 as functions of threading dislocation density, obtained from 004 rocking curve of coherently-strained 50 nm GaAs / 5 nm $In_{0.3}Ga_{0.7}As$ / 50 nm GaAs single quantum-well structure. The fringe intensities are normalized by the substrate peak intensity in order to eliminate the effect of incident intensity. The results indicate that each fringe width increases monotonically with thread density while the fringe intensity decreases monotonically with the thread density. The fringe widths are relatively constant below 5×10^5 cm^{-2} dislocations and then increase with the TDD On the other hand, the fringe intensities are relatively constant below 5×10^6 cm^{-2} TDD and increase with the TDD beyond that. The observed widths were also compared to those calculated on the basis of the convolution model. The widths predicted by the convolution model have correlated experimentally with known dislocation densities for single uniform layers, as given by [22]

$$\beta^2 \approx \beta_0^2 + \beta_\alpha^2 + \beta_\beta^2$$
$$= \beta_0^2 + 2\pi b^2 (TDD) \cdot ln(2) + 0.16\, b^2(TDD)\left|ln\left(2 \times 10^7\, cm\sqrt{(TDD)}\right)\right| tan^2\left(\theta_B\right) \quad (19)$$

where β_α and β_β respectively are the widths of the angle-scale distributions associated with angular mosaic spread and d-spacing mosaic spread. β_o is the intrinsic rocking curve width determined by using a dynamical simulation for a perfect crystal with the same thickness. These results are given in Fig. 7(a). The fringe widths calculated by the mosaic crystal model are almost in agreement with the widths predicted by the convolution model. The threading dislocation density in a single quantum well may be estimated by either approach therefore.

The threading dislocation density can also be characterized in the quantum-well layer by using the width or normalized intensity of the primary Bragg peak for the quantum well. Figures 8 and 9 display the calculated 113 and 224 rocking curve widths (a) and intensities (b) from a coherently-strained 50 nm GaAs / 5 nm In₀.₃Ga₀.₇As / 50 nm GaAs single quantum-well on a GaAs (001) substrate, subjected to thread defect density

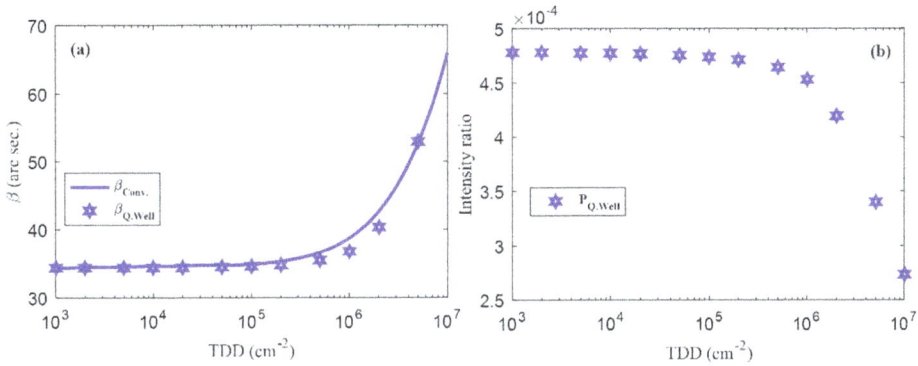

Fig. 8. Calculated 113 rocking curve widths (a) and intensity ratios (b) for the primary Bragg peak of the quantum well, as a function of the threading dislocation density (TDD). In (a), the width is compared to the convolution model, as indicated in the key.

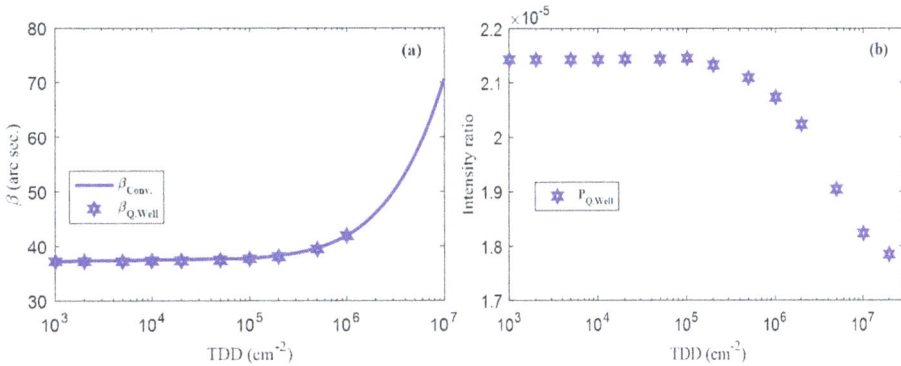

Fig. 9. Calculated 224 rocking curve widths (a) and intensity ratios (b) for the primary Bragg peak of the quantum well, as functions of the threading dislocation density (TDD). In (a), the width is compared to the convolution model as indicated in the key.

variations. The intensity ratio decreases monotonically with the dislocation density above a density of 10^5 cm^{-2} in both cases. While it is also true that the width of the primary peak increases monotonically with the TDD in both cases, it is less sensitive than the normalized intensity. The normalized intensity could be used to detect metamorphic growth if diffraction profiles were measured for several samples having different quantum well thicknesses, by plotting the normalized intensity as a function of thickness. An abrupt drop in intensity would accompany the increase in the threading dislocation density.

4. Conclusion

We have investigated the variation of several hkl rocking curves with dislocation density for a coherently-strained GaAs/In$_{0.3}$Ga$_{0.7}$As/GaAs single quantum well structure using diffraction profiles calculated using the mosaic crystal model. We show that the widths or normalized intensities of Pendellösung fringes or the primary Bragg peak may be used to characterize the dislocation density, and therefore the pseudomorphic-metamorphic transition, in a single quantum well.

Acknowledgement

The work has been supported in part by Taif University and Epitax Engineering. This support is gratefully acknowledged.

References

1. I. J. Fritz, T. J. Drummond, G. C. Osbourn, J. E. Schirber, and E. D. Jones, App. Phys. lett., 48, 1678 (1986).
2. H. Wang, H. Yu, X. Zhou, Q. Kan, L. Yuan, W. Chen, D. Ying, J. Pan, App. Phys. lett., 105, 141101 (2014).
3. R. Droopad, R. A. Puechner, K. T. Shiralagi, K. Y. Choi, and G. N. Maracas, App. Phys. lett., 58, 1777 (1991).
4. A. Ulyanenkov, T. Baumbach, N. Darowski, U. Pietsch, K. H. Wang, A. Forchel, and T. Wiebach, J. App. Phys., 85, 1524 (1999).
5. L. C. Hirst, R. J. Walters, M. F. Fuher, and N. J. Ekins-Daukes, App. Phys. lett., 104, 231115 (2014).
6. V. S. Speriosu, J. Appl. Phys., 52, 6094 (1981).
7. V. S. Speriosu and T. Vreeland Jr., J. Appl. Phys., 56, 1591 (1984).
8. L. Tapfer and K. Ploog, Phys. Rev. B, 40, 9802 (1989).
9. C. R. Wie, J. Appl. Phys., 65, 1036 (1989).
10. C. R. Wie and H. M. Kim, J. Appl. Phys., 69, 6406 (1991).
11. S. Takagi, Acta Cryst., 15, 1311 (1962).
12. D. Taupin, C. R. Acad. Sci., 256, 4881 (1963).
13. S. Takagi, J. Phys. Soc. Jpn., 26, 1239 (1969).
14. M. A. G. Halliwell, M. H. Lyons, and M. J. Hill, J. Cryst. Growth, 68, 523 (1984).
15. C. R. Wie, T. T. Tombrello, and T. Vreeland Jr., J. Appl. Phys., 59, 3743 (1986).
16. W. J. Bartels, J. Hornstra, and D. J. W. Lobeek, Acta Cryst., A42, 539 (1986)
17. P. B. Rago and J. E. Ayers, J. Electron. Mater., 42, 2450 (2013).

18. F. A. Althowibi, P. B. Rago, and J. E. Ayers, J. Vac. and Sci. Technol. B, 34, 041209 (2016).
19. Prins J A, Z. Phys., 63, 477 (1930).
20. P. Gay, P. B. Hirsch, and A. Kelly, Acta Met., 1, 315 (1953).
21. M. J. Hordon and B. L. Averbach, Acta Met., 9, 237 (1961).
22. J. E. Ayers, J. Cryst. Growth, 135, 71 (1994).
23. P. B. Rago and J. E. Ayers, Connect. Symp. on Microelect. and Optoelect., Storrs, CT (April 9, 2014).
24. F. A. Althowibi, and J. E. Ayers, IJHSES, 26, 1740011(2017).
25. J. E. Ayers, T. Kujofsa, P. Rago, and J. E. Raphael, Heteroepitaxy of Semiconductors: Theory, Growth and Characterization, Second Edition, 2016.
26. J. A. Ibers and W. C. Hamilton (Eds.), International Tables for X-ray Crystallography: Revised and Supplementary Tables: Kynoch Press for the International Union of Crystallography, 1974.

The Measurement of Microwave Absorption Characteristics of Nanocomposites Using a Coaxial Line Technique

Ahmet Teber*, Rajeev Bansal[1]

*Electrical and Computer Engineering, University of Connecticut, 371 Fairfield Way,
U-4157, Storrs, CT, 06269-4157, USA*
ahmet.teber@uconn.edu, [1]rajeev.bansal@uconn.edu

Ibrahim S. Unver[2], Zilhicce Mehmedi[3]

*[2]Department of Physics, Gebze Technical University, Microwave Lab.,
Gebze Technical University, Gebze, Kocaeli, 41400, Turkey*
*[3]Polymer Science and Technology Department, Istanbul Technical University,
ITU Ayazaga Campus, Maslak, Istanbul, 34496, Turkey*
[2]iunver@gtu.edu.tr, [3]mehmedi15@itu.edu.tr

Microwave absorption properties (MAP) of manganese soft spinel ferrite ($MnFe_2O_4$) nanoparticles (MSF NPs) mixed with multi-walled carbon nanotubes (MWCNTs) and molded as toroid-shaped pellets were experimentally studied using a coaxial line technique in the frequency range of 2-18 GHz. We used the coaxial line technique because of the smaller sample size, a wider frequency range, and a uniform cross section. The MAP were derived from the measured constitutive parameters, according to the transmission line theory. The results indicate that MWCNTs blended with ferrite nanoparticles represent a promising choice for broadband microwave absorbing materials.

Keywords: Dielectric and magnetic properties; radar absorbing materials; composites.

1. Introduction

It has been reported that, in applications requiring radar absorbing materials (RAM), both dielectric and magnetic materials have relatively low microwave absorption when they are used by themselves [1, 2]. It is possible to enhance absorption characteristics when magnetic powder nanoparticles are blended with dielectric nanomaterials [3]. For microwave absorption properties measurements, there exist several different techniques such as waveguide, coaxial line, resonant cavity, and free-space [4-7]. In this study, we have adopted a reflection/transmission technique in a coaxial line because of the smaller sample size, a wider frequency range, and a uniform sample cross section. The coaxial line is especially suitable for solid (powder) materials. The method is also preferable for

---

[1]Corresponding author.

measurements on powder samples because it uses less paraffin in the sample than the waveguide method [3].

2. Fabrication and Measurement Setup

2.1. *Preparation of manganese spinel ferrite nanoparticles (MSF NPs) and the other additives of composites*

Manganese and Zinc ferrite particles in the present study were prepared by a citric acid assisted sol-gel method using a procedure reported by Demir *et al.* [8] and Teber *et al.* [3]. The steps of the sol-gel process are the formation of the homogeneous precursor solution, sol, gel, aging, desiccation, and densification. MWCNTs were used as dielectric materials in order to create conductive paths within composite samples. The MWCNTs were purchased from US Research Nanomaterials, Inc., USA.

2.2. *Fabrication of nanocomposite samples*

For coaxial line measurements, the samples have to have a toroidal shape. The material must fit tightly in the sample holder in order to reduce the measurement uncertainty caused by air gaps. We designed our own manufacturing setup as shown in Fig. 1a.

The dielectric and magnetic nanoparticles were blended in a crucible with paraffin, and then the process was repeated for the molten blend with the help of a magnetic stirrer, which includes a heating plate and uses ultrasound waves to agitate particles. The molten blends were uniformly dispersed by stirring at 1000 rpm for 30 seconds. In the next step,

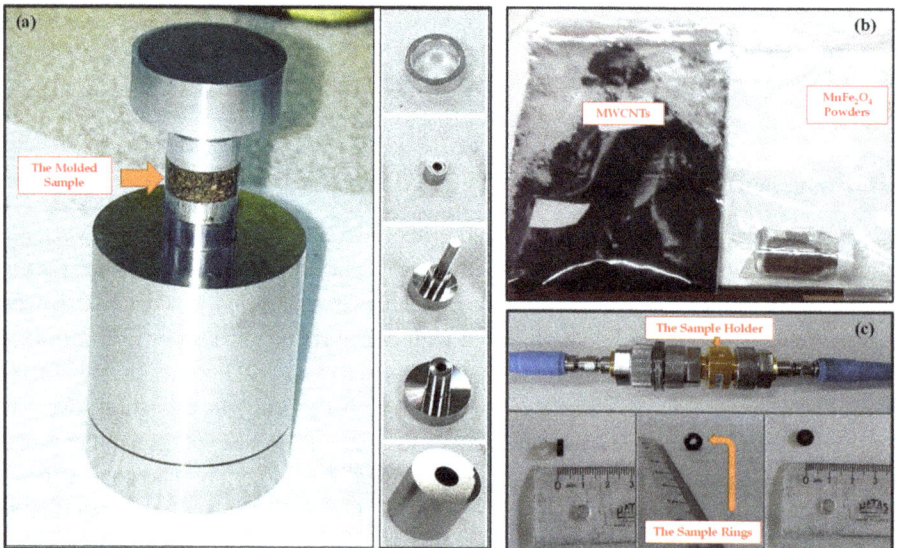

Fig. 1. (a) The manufacturing mold system; (b) The powders of MWCNTs, manganese spinel ferrites spinel ferrites; (c) The toroidal shape of the samples.

the molten blend was poured into the sample holders, which are the same size as a coaxial line cross section. The blends in the sample holder were pre-cured at room temperature for half an hour. For the last step, the sample holders including the pre-cured blends were flattened with a presser to avoid any air gaps in the coaxial line system. The composite samples were fabricated into toroid-shaped samples with an inner diameter of 3.5 mm, an outer diameter of 7 mm, and a thickness of 3 mm. A gold plated coaxial airline with a precision 7 mm connector was used to hold the sample [9, 10]. Then, the samples were ready for the microwave measurements of Fig. 1c.

Three different samples (MWCNTs-based magnetic nanocomposites of MSF NPs) were fabricated by incorporating in a proportion of 80 wt. percentage (the total filling amount) into 20 wt. percentage of a host matrix of molten paraffin. The filler materials (ferrite particles and MWCNTs) were blended in various mass proportions with the steps of 25-50-75%.

2.3. *The method of microwave measurements*

2.3.1. *Coaxial line measurements*

The complex scattering parameters corresponding to reflection (S_{11}) and transmission (S_{21}) parameters of a transverse electromagnetic (TEM) wave were measured using a vector network analyzer (Agilent PNAE8364B) in conjunction with a coaxial line system (Keysight Tech. 85050C 7mm coaxial connectors) incorporating samples of MSF NPs blended with MWCNTs for the frequency range of 2-18 GHz. Then, the electromagnetic constitutive parameters (ε and μ) were extracted from s-parameters using the material measurement software of 85071E (Keysight Tech.). In addition, the return loss (RL) of the samples was estimated from the computed complex permittivity and permeability values in order to assess their suitability as microwave absorbing materials.

Figure 2 indicates that the sample is placed between two parts of the coaxial line using a sample holder.

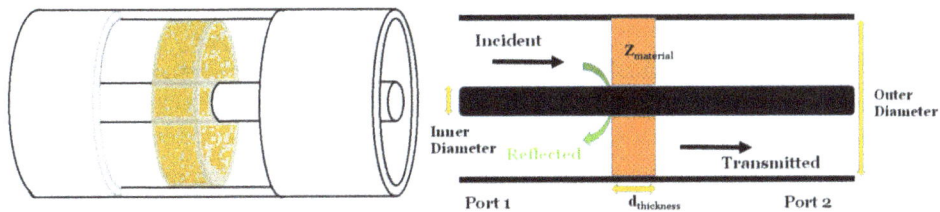

Fig. 2. The Measurement schematic of (a) Coaxial line and sample; (b) the side view of measurement setup.

2.3.2. *The computation of return loss*

The microwave absorption properties of the absorber samples can be defined by the reflection loss (RL). The RL of a ground-plane-backed samples can be calculated from the measured values of complex relative permittivity and permeability for the given frequency and absorber thickness using the following equations [3]:

$$RL(dB) = 20\log\left|\frac{(Z_{in}-1)}{(Z_{in}+1)}\right| \qquad (2)$$

while the normalized input impedance (Zin) is calculated by:

$$Z_{in} = \sqrt{\frac{\mu_r}{\varepsilon_r}} \tanh\left(-j\frac{2\pi fd}{c}\sqrt{\mu_r\varepsilon_r}\right) \qquad (3)$$

where f is the microwave frequency, d is the thickness of the sample, and c is the velocity of electromagnetic wave in a vacuum. ε_r and μ_r are the complex relative permittivity and permeability, respectively, which can be calculated by ε_r (= $\varepsilon'- j\varepsilon''$) and μ_r (= $\mu'- j\mu''$).

3. Results and Discussion

3.1. *The structure and morphology of the samples and the Energy Dispersive X-Ray (EDX)*

The crystal phases of the resultant composites were investigated using the model of Rigaku D/Max-3C using CuKα radiation (Rigaku Innovative Technologies, Inc., Germany) in the 2θ range of 20-80⁰ with a step size of 0.1 (Fig. 3).

Figure 4 shows the SEM images of the nano-structured samples, which demonstrates the dispersion of MWCNTs particles and $MnFe_2O_4$ NPs.

The morphology features of the nanocomposites were characterized by scanning electron microscopy (SEM) with the model of JSM-7001F SEM-EDS-EBL, JOEL USA, Inc., MA, USA, that operated with an accelerating voltage of 5kV as well as the different magnification from 100x (200µm) to 2000x (10µm). SEM was examined observing the amount of deposited filler nanoparticles of manganese spinel ferrites and MWCNTs. The fillers were dispersed randomly in the host matrix. In addition, it is clearly seen that some of MWCNTs were coated/filled with MSF NPs. This result agrees with the XRD results, where the absence of any additional peaks implies that MWCNTs were coated with spinel ferrite nanoparticles.

Figure 5 shows the EDX technique supplies the effective atomic concentration of different constituents in the top surface layers of solids. The energy dispersive X-ray (EDX) analysis of the specimens was carried out to confirm the elemental composition of the specimens. Energy-dispersive X-Ray spectroscopy (EDX) measurements were employed at 10 keV with INCA Instruments (Oxford Inca System, Oxford Co. Ltd., UK). The EDX spectrum confirms the presence of Mn, Fe, C, and O elementals for the samples of $MnFe_2O_4$ with MWCNTs. The other minor peaks might be due to the impurities in the starting materials. The relative atomic and weight abundance of Mn, Fe, and Oxygen species are presented in the table on EDX graph. The atomic ratios of Fe/Mn are found to be 2:1 by EDX analyses, confirmed the presence of $MnFe_2O_4$ nanoparticles on the MWCNT surfaces.

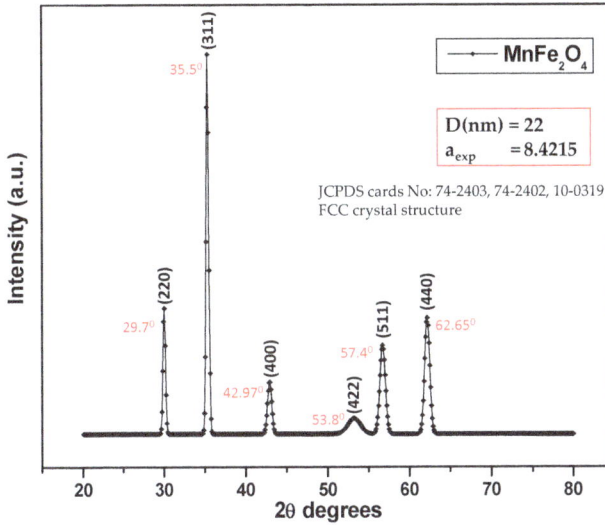

Fig. 3. XRD patterns of samples fabricated with MZF NPs by blending MWCNTs.

Fig. 4. The Surface Morphology of the specimens ($MnFe_2O_4$ blended with MWCNTs).

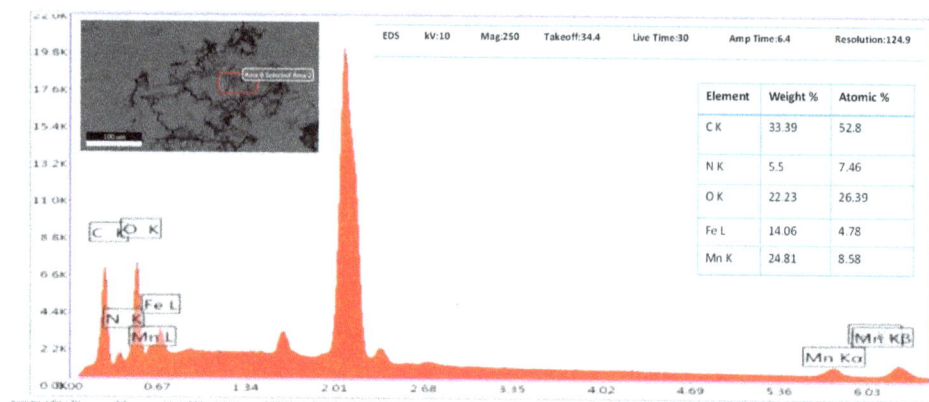

Fig. 5. The energy dispersive X-ray (EDX) spectrum of $MnFe_2O_4$ magnetic nanoparticles on multi-walled carbon nanotubes.

3.2. *The electromagnetic properties of composite samples*

The electromagnetic constitutive parameters were extracted using an Agilent PNA E8364B network analyzer (Keysight Tech.) in conjunction with a coaxial line system (Keysight Tech. 85050C 7 mm coaxial connectors) at 2.4-18 GHz frequency range at room temperature. The real (ε_r' and μ_r') and the imaginary part (ε_r'' and μ_r'') of relative complex permittivity and permeability as well as the dielectric loss tangent (tan (δ_ε) $=\varepsilon_r''/\varepsilon_r'$) and the magnetic loss tangent (tan (δ_μ) $=\mu_r''/ \mu_r'$) for nano-structured composites were extracted with the help of the software (85071E Keysight Tech.), pre-installed in the network analyzer. The real part of relative complex permittivity and permeability are associated with the storage capability of electric and magnetic energies. The ε'' and μ'' are associated with energy loss within materials, resulting from dielectric and magnetic mechanisms such as conduction, resonance, interfaces, relaxation and atomic, electronic and/or dominant polarization.

The relative permittivity real part (ε'), the relative permeability real part (μ'), the dielectric loss tangent (tan $\delta\varepsilon = \varepsilon''/\varepsilon'$), and the magnetic loss tangent (tan $\delta\mu = \mu''/\mu'$) of manganese spinel ferrites blended with MWCNTs with different weight percentages are presented in Fig. 6. The weight percentages are namely in 25% increments of weight ratios (Table 1).

Table 1. The proportion used for the sample preparation.

Sample Code	Proportions
1	75 wt % (MWCNTs) with 25 wt % ($MnFe_2O_4$)
2	50 wt % (MWCNTs) with 50 wt % ($MnFe_2O_4$)
3	25 wt % (MWCNTs) with 75 wt % ($MnFe_2O_4$)

In Fig. 6a and b, the ε' values of the samples slightly decrease in the range of 10.4-1.1. The dielectric tangent loss slightly decreases in the range of 0.22-0.04 within the frequency range 2.0-8.3 GHz. It increases in the range of 0.04-0.16 within the frequency range 8.3-12.5 GHz. The dielectric tangent loss of the samples in the rest of the frequency range exhibits almost a constant value with only a small fluctuation. The constancy in the values of the real part of permittivity suggests that there was a dominant polarization, in which the oscillation of the electric dipole moments was in phase or somewhat out of phase with the microwave frequencies [10].

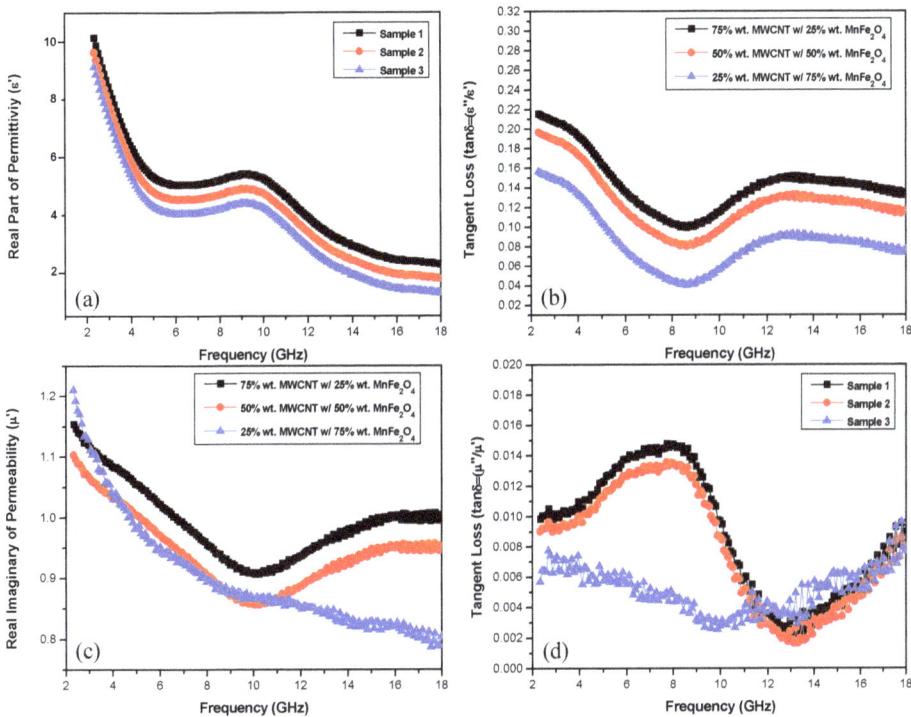

Fig. 6. The electromagnetic constitutive parameters (a) the relative real part of permittivity (b) The Dielectric tangent loss of the samples, (c) the relative imaginary part of permeability (d) The Magnetic tangent loss of the samples, resulting from $MnFe_2O_4$ magnetic nanoparticles on multi-walled carbon nanotubes.

As demonstrated in Fig. 6c and d, we can observe that the μ' values of the samples slightly decreased in the frequency range of 2-10 GHz, whereas its values increased between 10-18 GHz. The μ'' values (or tan $\delta\mu$) of the samples indicate broad resonance peak in the whole frequency range, which is related to the small size effect, surface effect, and spin wave excitations [11].

It is observed that the tan $\delta\varepsilon$ values are higher than tan $\delta\mu$, suggesting that the magnetic loss is low. MWCNTs are often used as an additive coating with manganese spinel ferrites to increase its conductivity [12]. According to the free-electron theory, increased conductivity of a composite could result in strong dielectric loss, and thus, the blend of MWCNTs and MSF NPs show a stronger dielectric loss in 2-18 GHz.

3.3. *Microwave absorption properties*

First, the results obtained by the previous waveguide [1 and the present coaxial-line methods over the common frequency range (8-12 GHz) are compared below in Fig. 7 and Table 2 and show good agreement.

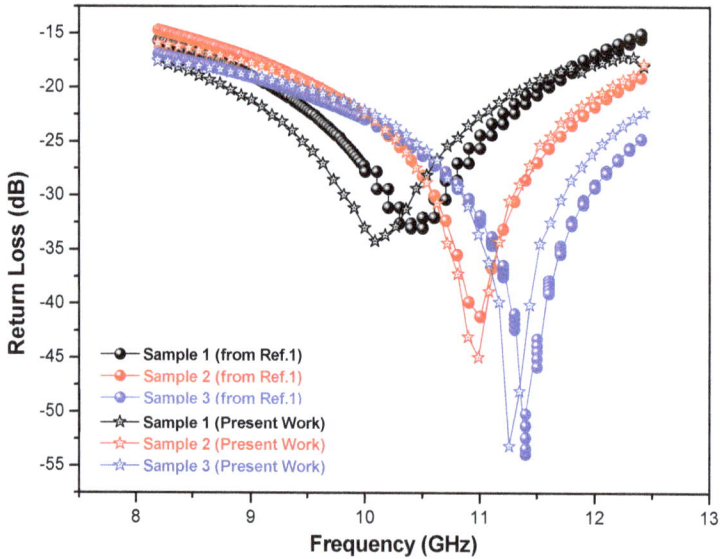

Fig. 7. Comparison Return Losses of Coaxial Line and Waveguide Measurements in the frequency range of 8-12 GHz.

Table 2. Microwave absorption results.

Code	f_r (GHz)		Return Loss (dB)		Bandwidth (under −20 dB)	
	Present Work	Previous Work	Present Work	Previous Work	Present Work	Previous Work
1	10.08	10.44	−34.07	-32.84	1.98 GHz	1.88 GHz
2	11.00	11.00	−45.16	-41.20	2.55 GHz	2.20 GHz
3	11.26	11.41	−53.31	-56.00	3.49 GHz	3.38 GHz

In Fig. 8, we present the calculated return loss (RL) of the samples for the various compositions of additives over the frequency range 2-18 GHz. The samples thickness are approximately the same thickness (3 mm). It can be observed that the plots of RL versus frequency indicate good microwave absorption. The minimum RL values shifted to higher frequencies and the bandwidths increased with an increasing amount of manganese ferrite material in the mixture. The lowest RL value and the bandwidth change with different loadings of manganese and zinc spinel ferrite nanoparticles. A minimum RL of -53.31 dB at 11.26 GHz with a bandwidth of 3.49 GHz (RL < -20 dB) was obtained in MWCNTs with $MnFe_2O_4$ NPs ($x = 0.0$).

These comparison results are summarized in Table 2, including the sample codes, the center frequencies (f_r), the values of minimum achievable return loss (dB), and the corresponding bandwidths (GHz).

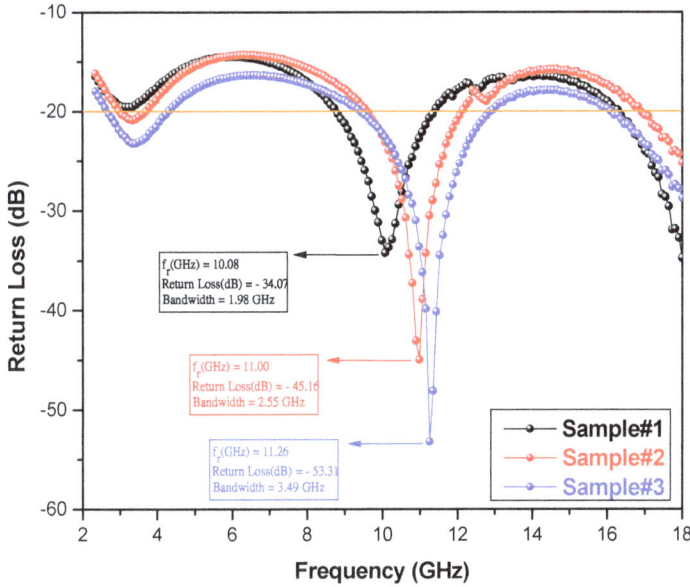

Fig. 8. Return Loss versus frequency of MWCNTs with MSF NPs alloy with Coaxial Line Technique.

4. Conclusion

Because of both dielectric and magnetic contributions to microwave absorption properties at microwave frequencies, the composite samples exhibited desirable microwave absorption properties in terms of the minimum RL values and the broad absorption bandwidths. The minimum RL value tends to shift to higher frequencies with an increasing amount of manganese ferrite material in the mixture. These results support a promising role for MWCNTs blended with ferrite nanoparticles for broadband microwave absorbing materials.

Acknowledgements

A. Teber is grateful to the Ministry of National Education of the Republic of Turkey (Grant number: YLSY-2009) for his graduate fellowship support.

References

1. Z. Wang, L. Wu, J. Zho, W. Cai, B. Shen, and Z. Jiang, *The Journal of Physical Chemistry C*, **117**, 5446-5452 (2013).
2. T. Giannakopoulou, L. Kompotiatis, A. Kontogeorgakos, and G. Kordas, *J. Magn. Magn. Mater.* **246**, 360-365 (2002).
3. A. Teber, K. Cil, T. Yilmaz, B. Eraslan, D. Uysal, G. Surucu, A. Baykal, and R. Bansal, *Aerospace.* **4(1)**, 2, (2017).
4. K. Y. You, *Microwave Systems and Applications*, eds. K. G. Sotirios (InTech, 2017).
5. A. Ozturk, R. Suleymanli, B. Aktas, and A. Teber, *Chinese Physics Letters.* **29(2)**, 027301 (2012).
6. A. Ozturk, R. Suleymanli, B. Aktas, and A. Teber, in *Microwave Symposium (MMS)*, Mediterranean, **2010**, p. 4-7, (IEEE).
7. A. Teber, I. Unver, H. Kavas, B. Aktas, and R. Bansal, *J. Magn. Magn. Mater.* **406**, 228-232 (2016).
8. A. Demir, S. Guner, Y. Bakis, S. Esir, and A. Baykal, *J. Inorg. Organomet. Polym. Mater.* **24**, 729-736 (2014).
9. Y. Qing, W. Zhou, F. Luo, and D. Zhu, String field theory, *Carbon.* **48** 4074-4080 (2010).
10. S. M. Abbas, A. K. Dixit, R. Chatterjee, and T. C. Goel, *Journal of Magn. Magn. Mater*, **309(1)**, 20-24 (2007).
11. P. Liu, Y. Huang, J. Yan, Y. Yang, and Y. Zhao, *ACS Applied Mater. & Interfaces*, **8(8)**, 5536-5546 (2016).
12. M. Z. Wu, Y. D. Zhang, S. Hui, T. D. Xiao, S. H. Ge, W. A. Hines, J. I. Budnick, and G. W. Taylor, *Applied Physics Letters*, **80**, 4404 (2002).

Additively Manufactured Inkjet-/3D-/4D-Printed Wireless Sensors Modules

Ajibayo Adeyeye[*], Aline Eid, Jimmy Hester, Tong-Hong Lin, Syed Abdullah Nauroze,
Bijan Tehrani and Manos M. Tentzris

*Department of Electrical and Computer Engineering, Georgia Institute of Technology,
Atlanta, GA 330332-0250, USA*
adeyeyao@gatech.edu

This publication considers the use of a variety of additive manufacturing techniques in the development of wireless modules and sensors. The opportunities and advantages of these manufacturing techniques are explored from an application point of view. We discuss first the origami (4D-printed) structures which take advantage of the ability to alter the shape of the inkjet-printed conductive traces on the paper substrate to produce a reconfigurable behavior. Next, focus is shifted towards the use of additive manufacturing technology to develop skin-like flexible electrical system for wireless sensing applications. We then discuss the development of a fully flexible energy autonomous body area network for autonomous sensing applications, the system is fabricated using 3D and inkjet printing techniques. Finally, an integration of inkjet and 3D printing for the realization of efficient mm-wave 3D interconnects up to 60GHz is discussed.

Keywords: Additive manufacturing; flexible electronics; inkjet printing; 3-D printing; energy harvesting; modules; origami; wireless sensors; Van Atta Array; smart skins.

1. Introduction

There is an ever-increasing need for a more intelligent and interactive environment that makes use of wireless sensor-based technologies including large scale sensor networks, the Internet of Things (IoT) and smart skins (SS). With the large number of devices required to satisfy this need and the costs associated with fabricating such a large amount, there is a high demand for low cost wireless sensors modules. Current manufacturing techniques are primarily subtractive and as a result there is a large amount of material wasted making this technique very expensive and largely unsustainable for the development of these devices. A potential solution to this problem is through the use of additive manufacturing technologies (AMT) such as inkjet, 3D and 4D printing techniques. AMTs have gained a lot of attraction over the past few years as they come with some very desirable qualities. They have much lower setup costs, produce easily repeatable results and have a much lower environmental impact when compared to subtractive techniques such as lithography.

---

*Corresponding author.

AMTs involve directly depositing different materials layer-by-layer so that we are able to form complex 3D structures while minimizing the amount of material wasted. AMTs are able to realize features as small as 1μm making them highly suitable for the development of wireless sensor modules for millimeter-wave applications and smart skins [1, 2].

This paper presents an overview of the use of AMTs in the development of wireless sensor modules from an application perspective. Section 2 discusses taking advantage of the flexibility of printed electronics to develop origami-inspired RF structures that produce wideband tunability by simply altering their shape. Section 3 investigates the use of AMTs to develop flexible and wearable skin-like electrical systems that are suitable for long range wireless sensing applications. Section 4 proposes a long-range energy autonomous wearable sensor system, fabricated using a combination of inkjet and 3D printing technologies that is able to achieve an order of magnitude increase in performance compared to conventional systems. Section 5 discusses the development of low loss 3D RF interconnects pertinent to the development of highly integrated wireless systems. The paper is concluded and summarized in Section 6.

2. Origami-Based Tunable RF Structures

Modern wireless and communication systems have multiple communication systems that operate on different frequency bands. Therefore, they require RF structures can tune their frequency response on-demand that fewer RF modules are required to for the whole system and reduce its overall size and cost. Some of the conventional methods to realize tunable RF structures include integration of complex or non-linear electrical structures, use of specialized substrates or integrated MEMS structures. However, these techniques become expensive and laborious as the overall size of the RF structure increases.

Recently, origami-inspired RF structures [1, 3-5] have attracted a lot of attention as an alternative methodology to realize wide-band tunability by simply changing its shape. One of the key limitations in implementation of such structures is realization of flexible conductive traces that do not crack or break when bent. Traditionally, this was achieved by using copper tape which is prone to peel off with environmental changes and makes fabrication process arduous, non-repeatable and unsuitable for high-frequency applications. In contrast, inkjet-printed conductive traces on cellulose paper presented in [6] do not suffer from these disadvantages as it uses the high porosity of the paper to embed conductive silver Nano-particles into the substrate. This allows conductive traces to bend along the substrate and maintain good conductivity for small radius of curvature. The flexibility can be improved further by using special "bridge-like" structures along the conductive traces [7].

This section presents an origami-inspired "shape-changing" 4D frequency selective surface (FSS) with dipole resonators that can tune its response as the structure is folded. This is realized by inkjet printing dipole resonators along the fold-lines of a Miura-Ori tessellation as shown in Fig. 1. The unit cell of Miura-Ori tessellation consists of four parallelograms (with lengths l_1 and l_2 and internal angle α) joined together along the edges.

One of the key advantages of Miura-Ori tessellation is that it has single degree of freedom, that is, its kinematics can be completely defined by variation in dihedral (fold) angle θ. Thus, as the value of θ is decreased from 180°, that is, completely flat configuration, the effective length of the dipoles and their inter-element distances decreases systematically. Thereby resulting in resonant frequency to shift to higher values as shown in Fig. 1(b). Moreover, it features high angle of incidence rejection [7] as a result of morphing of 2D flat dipoles into 3D V-shaped dipoles as the folding angle is decreased.

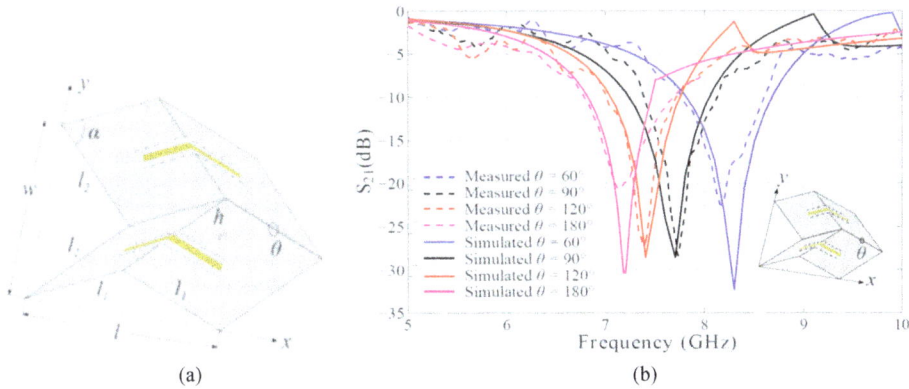

(a)   (b)

Fig. 1. (a) Unit cell of a Miura-FSS structure (b) measured and simulated results of Miura-FSS with different values of folding angle (θ).

3. Additive Manufacturing Technologies Bringing about the Internet of Skins

Countless research efforts are currently striving towards the development of novel technological approaches to create ubiquitous networks of objects, interfaced in a dense, intricate and ad-hoc network of "things", according to the expanding and now rather all-encompassing Internet of Things (IoT) concept. Most of these venture down the path of, on the one hand, the analysis and improvement of systems using currently available technologies through the implementation of optimized sensor measurement, data processing and communication schemes or, on the other hand, the lowering of the operating voltage and power consumption of the elements that now constitute standard IoT sensing nodes (motes): such as sensors, processors, and sub-2.4 GHz transceivers.

A tangential but distinct area of research, Flexible Hybrid Electronics (FHE), is now at the center of the development of manufacturing techniques enabling the creation of flexible and wearable skin-like electrical systems. Nevertheless, the interface between those two fields of work (IoT+FHE) is fraught with challenges from an electromagnetics standpoint, as flexible antennae operating in the frequency ranges generally trodden by IoT wireless schemes effectively have their performance compromised by the proximity of their mounting host.

This limitation can be overcome by the use of higher-frequency communications, which, for identical communication approaches, increases node complexity, cost, and power consumption. However, it has been shown that the use of RFID approaches at higher operating frequencies can not only enable the design and fabrication of mounting-insensitive skin-like devices to enable an "Internet of Skins", but can additionally enhance their performance and offer richer features, relative to their identically-sized lower-frequency counterparts. One of these skins was reported [8, 9] in the form of a chipless humidity sensing node leveraging the remarkable properties of a cross-polarizing Van-Atta retrodirective array structure, and operating in the 30 GHz to 40 GHz frequency range. The chipless RFID (shown in Fig. 2(a)) device not possessed a skin-like form factor but also was identified and radially located at a 30m range (shown in Fig. 2(b)) more than one order of magnitude higher than the state of the art before it.

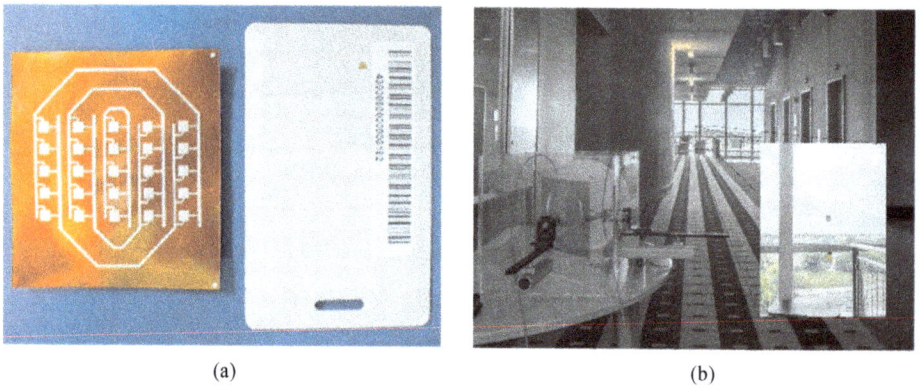

(a)   (b)

Fig. 2. (a) Chipless mm-wave RFID Smart Skin, (b) Interrogation of the chipless Smart Skin at 30m.

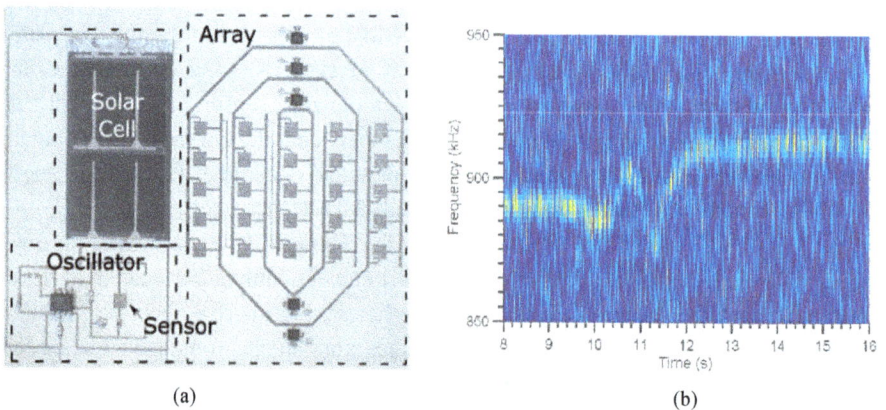

(a)   (b)

Fig. 3. (a) Mm-wave backscatter RFID ammonia sensing node, (b) Real-time ammonia exposure sensing data measured by the "Smart Skin".

While chipless approaches dominate the absolute lower bound of per-device cost, these remain relatively low performance and are, by nature, very features-limited. However, the same electromagnetics approach exploited in the aforementioned chipless prototypes can also be slightly altered to enable non-linear communications by means of nanowatt backscatter modulation schemes, as was reported in [10]. Doing so, fully-capable ultra-low-power and energy autonomous "Internet of Skins" devices, similar to the ammonia-sensing node shown on Fig. 3(a), can be constructed and mounted onto any surfaces to empower them with intelligence. Similar to its chipless counterpart, the reported device permitted environmental sensing (as shown on Fig. 3(b)) and displayed a reading range exceeding that of any of its contemporaneous backscatter communications art.

4. Long-Range Energy Autonomous Wearable Sensor System

A fundamental challenge in the design of autonomous sensing devices is the fact that they heavily rely on primary batteries that require replacement after a while. To overcome this, ambient energy harvesting is proposed to recharge energy storage devices, bistatic scatter radio, and chipless RFID.

Figure 4(a) and (b) shows a novel wearable and flexible energy autonomous on-body sensing network featuring full operability through energy harvesting from a hand-held

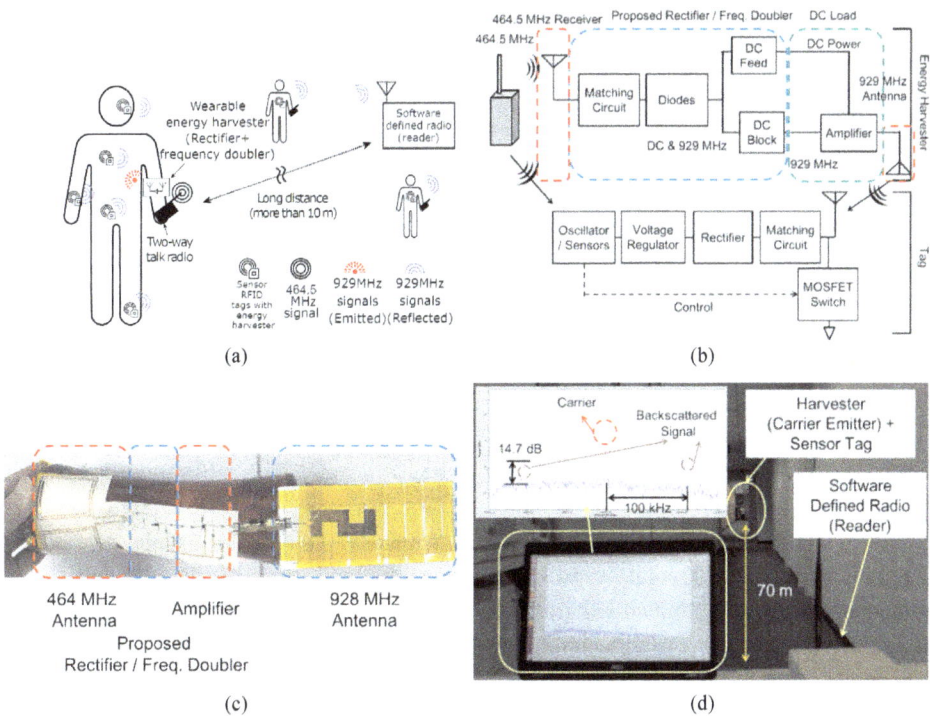

Fig. 4. (a) The system architecture, (b) system block diagram, (c) proposed energy harvester, and (d) field test of the inkjet and 3D-printed long-range energy autonomous wearable wireless sensor system [11].

464.5 MHz UHF two-way talk radio [11]. The two-way talk radio is used as the only energy source to generate all required power for the entire system. Furthermore, the system utilizes not only harvested DC power to power the ICs but also the harmonics generated during rectification as the carrier emitter to extend the reading ranges of the backscattered RFID tags. The system is fabricated using both inkjet printing and 3D printing techniques. The inkjet printing SU8 masks are used to pattern the circuits. The 3D printed substrates with desired thicknesses are used to fabricate the artificial magnetic conductor to shield the lossy effects of the human tissue. The proposed energy harvester is shown in Fig. 4(c). The field test of the proposed system is demonstrated in Fig. 4(d). The software defined radio is used as the RFID reader to detect the backscattered RFID tags. As depicted in Fig. 4(d), the backscattered RFID tags (sensors) located at 70 m away can be successfully detected by the reader with signal to noise ratio equals to 14.7 dB. Compared with the reading range of the conventional backscattered RFID tags which is 3 m, the proposed system can achieve 23 times longer reading range.

5. Fully-Printed 3D Ramp Interconnects

In a modern wireless system, efficient interconnects are necessary for interfacing various passive and active components within a single package in a highly-integrated system-in-package (SiP) design scenario. The design and implementation of these RF interconnects rely on the reduction of losses from impedance mismatch and material limitations along with the ability to realize 3D solutions for a variety of modern packaging scenarios. Wire and ribbon bond interconnects are typically used throughout industry as a simple and low-cost solution to surface mount and cavity-embedded RF devices, however issues of sway and high parasitic inductance are often present, requiring the inclusion of additional passive components to compensate at within the millimeter-wave (mm-wave) wireless regime [12]. Additive manufacturing techniques such as inkjet and 3D printing offer a variety of solutions for the realization of efficient RF interconnects necessary for the development of low-cost, highly-integrated wireless systems. First, the control of printed 3D topological profiles allows for the minimization of bonding wire loop length, resulting in the reduction of parasitic inductance and eliminating the need for passive compensation [13]. Second, the highly-reconfigurable patterning technologies used within additive manufacturing allow for the patterning of transmission line topologies typically limited to planar substrates, further reducing impedance mismatches from packaging substrate to device and minimizing losses.

Additionally, the development of printed sloped 3D interconnects allows for SiP integration reaching beyond a packaging substrate and towards the integration of components directly within an integrated circuit (IC) encapsulant. In order to evaluate the use of additive manufacturing technologies for 3D interconnection, a parametric study of 3D sloped interconnects is performed with a test vehicle fabricated through the combination of inkjet printing and 3D stereolithography (SLA) printing [14]. SLA printing is used to fabricate 3D slope structures with angles ranging from 15-75°. Next, inkjet printing is used with a silver nanoparticle-based ink to pattern 50 Ω coplanar waveguide

(CPW) interconnects, representing interconnection between a wireless IC device and a peripheral SiP component. Figure 5 shows the fully-printed 3D sloped interconnects for evaluation along with the measured and simulated insertion loss of the interconnects with various slopes. Slopes of up to 65° are achieved through the inkjet and 3D printing methods, where pattern and model modification have the potential to allow for higher gradient slopes. The measured and simulated insertion loss of the 3D sloped interconnects exhibit a maximum per-length loss of 0.5-0.6 dB/mm at 60 GHz, yielding a 10x reduction in insertion loss compared to comparable wire bond solutions at 60 GHz [15]. This demonstration of integrating 3D and inkjet printing technologies for efficient mm-wave 3D interconnects highlights the potential to realize fully-printed, highly-reconfigurable SiP wireless packages for emerging 5G and automotive radar applications in a truly additive fashion.

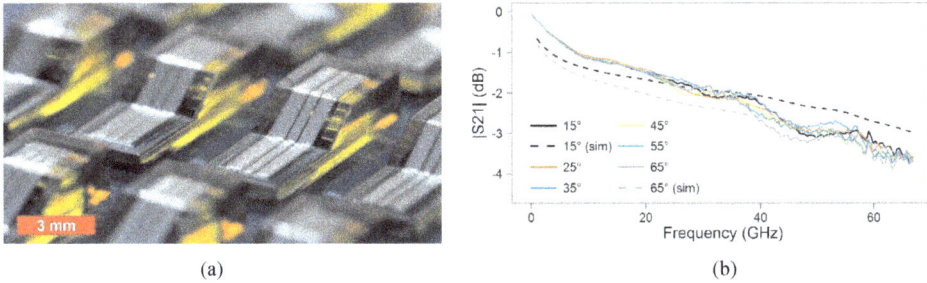

(a)   (b)

Fig. 5. Fully-printed 3D ramp interconnects: (a) CPW interconnects with 35° slope and (b) measured and simulated insertion loss of sloped interconnects ranging from 15-65° slope angles. [14]

6. Conclusion

This paper has explored various applications of additive manufacturing techniques in the development of wireless sensors and modules. We show the development of reconfigurable shape changing origami inspired RF structures used in the realization of tunable wideband frequency selective surfaces at microwave frequencies. In addition, the advantages of inkjet and 3D printing are highlighted in the discussion of the development of smart skins and energy autonomous sensors for long range wireless sensing. The applications discussed highlight the advantages that AMTs have over existing technologies for the development of wireless sensors and modules. AMTs show ease of scalability, repeatability and cost-effectiveness superior to techniques currently in use.

Acknowledgements

The authors would like to thank the National Science Foundation (NSF), Semiconductor Research Corporation (SRC), and the Defense Threat Reduction Agency (DTRA) for their support during this project.

References

1. S. A. Nauroze *et al.*, "Additively Manufactured RF Components and Modules: Toward Empowering the Birth of Cost-Efficient Dense and Ubiquitous IoT Implementations," in *Proceedings of the IEEE*, vol. 105, no. 4, pp. 702-722, April 2017.
2. Jo Bito, Ryan Bahr, Jimmy Hester, John Kimionis, Abdullah Nauroze, Wenjing Su, Bijan Tehrani, Manos M. Tentzeris, "Inkjet-/3D-/4D-printed autonomous wearable RF modules for biomonitoring, positioning and sensing applications," Proc. SPIE 10194, Micro- and Nanotechnology Sensors, Systems, and Applications IX, 101940Z (18 May 2017); doi: 10.1117/12.2262795; https://doi.org/10.1117/12.2262795
3. X. Liu, S. Yao, S. V. Georgakopoulos, B. S. Cook, and M. M. Tentzeris, "Reconfigurable helical antenna based on an origami structure for wireless communication system," in *IEEE MTT-S Int. Microw. Symp. Dig.*, Jun. 2014, pp. 1-4.
4. W. Su, S. A. Nauroze, B. Ryan and M. M. Tentzeris, "Novel 3D printed liquid-metal-alloy microfluidics-based zigzag and helical antennas for origami reconfigurable antenna "trees"," *2017 IEEE MTT-S International Microwave Symposium (IMS)*, Honolulu, HI, 2017, pp. 1579-1582.
5. X. Liu, S. Yao, B. S. Cook, M. M. Tentzeris and S. V. Georgakopoulos, "An Origami Reconfigurable Axial-Mode Bifilar Helical Antenna," in *IEEE Transactions on Antennas and Propagation*, vol. 63, no. 12, pp. 5897-5903, Dec. 2015.
6. S. A. Nauroze, J. Hester, W. Su and M. M. Tentzeris, "Inkjet-printed substrate integrated waveguides (SIW) with "drill-less" vias on paper substrates," *2016 IEEE MTT-S International Microwave Symposium (IMS)*, San Francisco, CA, 2016, pp. 1-4.
7. S. A. Nauroze, L. Novelino, M. M. Tentzeris and G. H. Paulino, "Inkjet-printed "4D" tunable spatial filters using on-demand foldable surfaces," *2017 IEEE MTT-S International Microwave Symposium (IMS)*, Honolulu, HI, 2017, pp. 1575-1578.
8. Hester, Jimmy GD, and Manos M. Tentzeris. "Inkjet-printed flexible mm-wave Van-Atta reflectarrays: A solution for ultralong-range dense multitag and multisensing chipless RFID implementations for IoT smart skins," *IEEE Transactions on Microwave Theory and Techniques* 64.12 (2016): 4763-4773.
9. Hester, J. G. D., and M. M. Tentzeris. "Inkjet-printed Van-Atta reflectarray sensors: A new paradigm for long-range chipless low cost ubiquitous Smart Skin sensors of the Internet of Things," *Microwave Symposium (IMS), 2016 IEEE MTT-S International*. IEEE, 2016.
10. Hester, Jimmy GD, and Manos M. Tentzeris. "A Mm-wave ultra-long-range energy-autonomous printed RFID-enabled van-atta wireless sensor: At the crossroads of 5G and IoT," *Microwave Symposium (IMS), 2017 IEEE MTT-S International*. IEEE, 2017.
11. T. H. Lin, J. Bito, J. G. D. Hester, J. Kimionis, R. A. Bahr and M. M. Tentzeris, "On-Body Long-Range Wireless Backscattering Sensing System Using Inkjet-/3-D-Printed Flexible Ambient RF Energy Harvesters Capable of Simultaneous DC and Harmonics Generation," in *IEEE Transactions on Microwave Theory and Techniques*, vol. 65, no. 12, pp. 5389-5400, Dec. 2017.
12. Y. P. Zhang and D. Liu, "Antenna-on-chip and antenna-in-package solutions to highly integrated millimeter-wave devices for wireless communications," *IEEE Transactions on Antennas and Propagation*, vol. 57, no. 10, pp. 2830-2841, 2009.

13. B. Tehrani, B. Cook, and M. Tentzeris, "Inkjet-printed 3d interconnects for millimeter-wave system-on-package solutions," *2016 IEEE MTT-S International Microwave Symposium (IMS)*, June 2016.

14. B. Tehrani, R. Bahr, W. Su, B. Cook, and M. Tentzeris, "E-band characterization of 3D-printed dielectrics for fully-printed millimeter-wave wireless system packaging," *2017 IEEE MTT-S International Microwave Symposium (IMS)*, June 2017.

15. T. Krems, W. Haydl, H. Massler, and J. Rudiger, "Millimeter-wave performance of chip interconnections using wire bonding and flip chip," *1996 IEEE MTT-S International Microwave Symposium Digest*, 1996.

All-Optical Logic Gates Based on Quantum-Dot Semiconductor Optical Amplifier

Xiang Zhang[*]

Department of Physics, University of Connecticut, 2152 Hillside Road, U-3046, Storrs, CT 06269, USA
xiang.zhang@uconn.edu

Sunil Thapa

Department of Physics, University of Connecticut, 2152 Hillside Road, U-3046, Storrs, CT 06269, USA
sunil.thapa@uconn.edu

Niloy K. Dutta

Department of Physics, University of Connecticut, 2152 Hillside Road, U-3046, Storrs, CT 06269, USA
nkd@phys.uconn.edu

We propose a scheme to realize all-optical logic operation in quantum-dot semiconductor optical amplifier (QD-SOA) based Mach-Zehnder interferometer (MZI) considering the effects of two-photon absorption (TPA). During propagation of sub-picosecond pulses in QD-SOA, TPA leads to an additional change in carrier recovery dynamics in quantum-dots. We utilize a rate equation model to take into account carrier refill through TPA and nonlinear dynamics including carrier heating and spectral hole burning in the QD-SOA. The simulation results show the TPA induced pumping in the QD-SOA can reduce the pattern effect and increase the output quality of the all-optical logic operation. With TPA, this scheme is suitable for high speed Boolean logic operation at 320 Gb/s.

Keywords: Two-photon absorption; optical logic; quantum-dot semiconductor optical amplifier.

1. Introduction

All-optical signal processing is expected to be important in future tera bit rate telecommunication networks [1]. All optical logic gates operating at high speeds will play important roles in future all optical networks, including signal regeneration, all optical packet routing and data encryption [2, 3]. Recently, there have been extensive research on various schemes of optical logical gates including traditional Boolean logic functions such as XOR, OR, AND, Set-Reset, etc. [4-14]. These schemes include using light beam interference in silicon photonic crystal [7], photonic crystal ring resonator [4], four-wave

[*]Corresponding author.

mixing in semiconductor optical amplifier (SOA) [6], binary phase shift keyed signal [13] and dual semiconductor optical amplifier Mach-Zehnder interferometer (SOA-MI) [8-10]. Among these schemes, the SOA based MZI has the advantage of being stable, compact, and simple, which has been demonstrated to be capable of all optical switching such as XOR, Set-Reset and flip-flop [8-11]. Especially with the introduce of SOA based on quantum-dot (QD) active region, the all optical switching speed reached ~250 Gb/s [8, 10]. These schemes will require high speed optical pulse train with ultra-short pulse width as the data carriers. Nowadays, various progresses have been made to generate ultra-short pulse at high repetition rate [15-18]. The QD excited state carriers can act as a reservoir which results in an ultrafast gain recovery. Berg *et al.* [19] claimed that QD-SOA does not support tera bit all optical logic operation due to the slow recovery of carriers in wetting layers. However, in their model, they assumed that the direct injection of carriers into wetting layer of a QD-SOA was only from the injected current. This is not the case during the propagation of sub-picosecond optical pulses because two-photon absorption (TPA) will generate carriers in the bulk region of a QD-SOA, which contributes to the carrier dynamics. By taking TPA into account, Ju *et al* demonstrated that the pattern effects are reduced when a train of optical pulses is injected in a QD-SOA, thus it is possible to operate at a higher speed [20]. Ju's simulation also has its limitations. It neglects gain saturation and nonlinear effects and is only based on inter-band carrier transitions. These effects dominate the device's gain dynamics above certain input power level.

In this work, we study the effects of TPA on optical logic operation based on QD-SOA-MZI. Rate equation approach has been widely used to simulate the carrier dynamics on QD-SOA systems assuming ultrafast dynamics. We present a model to consider wetting layer carriers refill through TPA as well as nonlinear effects affecting the gain and phase dynamics of the QD device. Our results show that with the consideration of TPA, optical logic gates based on QD-SOA-MZI have an improved output qualify and are capable of operating at 320 Gb/s.

2. All-Optical Logic Gates

Due to the compact and stable structure, SOA-MZI has been widely used in optical logic gates [21, 22]. Figure 1 presents a schematic diagram for QD-SOA-MZI based optical XOR gate. The principle of logic XOR operation utilizing the cross gain modulation (XGM) and cross phase modulation (XPM) processes in SOAs has been discussed and analyzed [9, 21]. The function of A XOR B is realized as follows. Data streams A and B are carried by two optical pulse streams at wavelength λ_1. They are injected into port 1 and 2, respectively. There is a clock stream centered at λ_2 injected into port 3 and it is evenly split into the two arms of the MZI. Initially, the MZI is unbalanced with a phase difference π between these two branches. As the two clock streams travel in the QD-SOAs, their phases and amplitudes are modulated. When they recombine at port 4, their interference will produce different results under different initial + conditions. For example, if the input A is '0' and B is '0', then these two clock streams experience the same gain and phase shift in QD-SOA and when they recombine at port 4, considering a phase difference π, they will undergo destructive interference and the output result is '0'. If A is '1' and B is '0', the gain and phase modulation of the two clock streams are different and their interference will produce '1'. This realizes the functionality of XOR.

We can also use this same scheme to realize the logic AND operation as showed in Fig. 2. By inputting data stream A and B into port 1 and 3, we will get a pattern of A AND B out from port 4. A low power CW light is normally injected into port 2 to cancel the background noise of data stream A out.

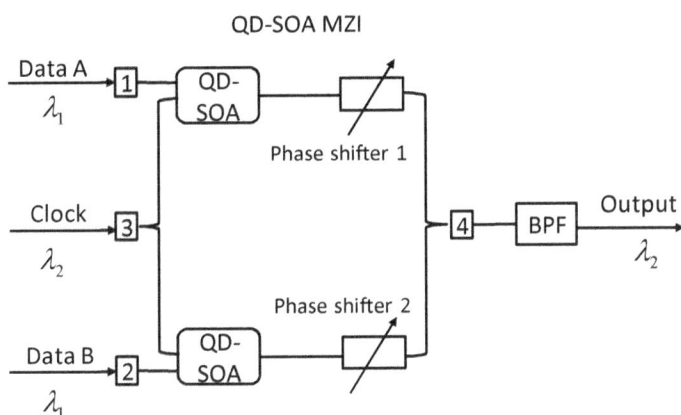

Fig. 1. Schematic diagram of the QD-SOA-MZI based XOR gate. BPF: bandpass filter centered at λ_2. Two phase shifters are used to induce a π phase difference in two arms.

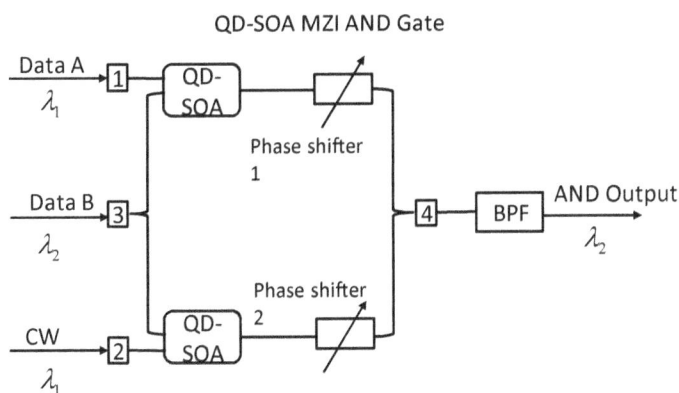

Fig. 2. Schematic diagram of the QD-SOA-MZI based AND gate. Data are inputted to port 1 and 3.

3. Working Principle

The device we choose here to construct the all-optical logic gate is the commonly discussed InAs/GaAs QD-SOA, in which InAs quantum dots are embedded in GaAs layers [23-25].

The gain of this type of device around 1.55 μm is typically ~15 dB and the noise figure is low at ~7 dB [23]. Also the gain of this device is nearly polarization independent [25]. Here we use the three-level rate equation model to simulate the carrier transitions in the device. Figure 3 illustrates the optical gain, the TPA process and carrier transition between the wetting layer (WL), the QD excited state (ES), and the QD ground state (GS). The device gain is determined by the carrier density of the QD ground state. The TPA generates carriers in the bulk region. These carriers then relax to the WL, and eventually are captured into QDs on ultrafast timescale [20]. In our model, we ignore the barrier dynamics and assume that carriers are injected directly from the contacts into the WL [19]. Since the only recipient of the pump current is the WL, and the QD excited state serves as a carrier reservoir for the ground state with ultra-fast carrier relaxation to the latter, the device gain dynamics is affected by their carrier densities and transition rates.

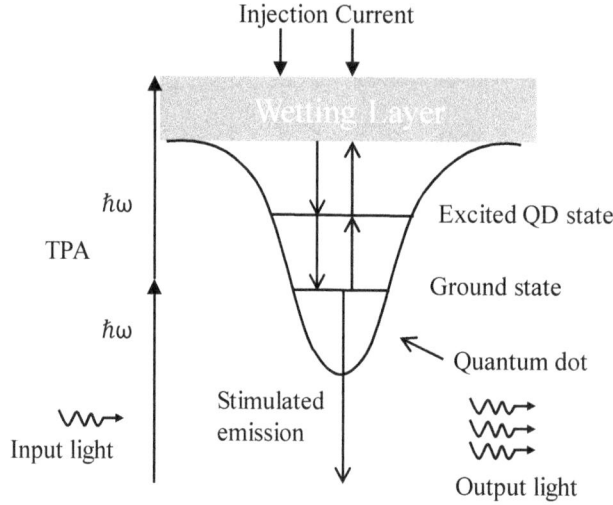

Fig. 3. A schematic of carrier dynamics with TPA in an InAs/GaAs QD-SOA. Quantum dots are embedded in the wetting layer.

The change in carrier densities of the three energy levels including the TPA process are described by the following coupled rate equations [14, 19, 20, 26]:

$$\frac{dw}{dt} = \frac{I}{eVN_{wm}} - \frac{w}{\tau_{wr}} - \frac{w}{\tau_{w-e}}(1-h) + \frac{N_{esm}}{N_{wm}}\frac{h}{\tau_{e-w}}(1-w) + \frac{\kappa}{2\hbar\omega N_{wm}}\left[\frac{S(t)}{A}\right]^2, \quad (1)$$

$$\frac{dh}{dt} = -\frac{h}{\tau_{esr}} - \frac{N_{wm}}{N_{esm}}\frac{w}{\tau_{w-e}}(1-h) - \frac{h}{\tau_{e-w}}(1-w) + \frac{N_{gsm}}{N_{esm}}\frac{f}{\tau_{g-e}}(1-h) - \frac{h}{\tau_{e-g}}(1-f), \quad (2)$$

$$\frac{df}{dt} = -\frac{f}{\tau_{gsr}} - \frac{f}{\tau_{g-e}}(1-h) + \frac{N_{esm}}{N_{gsm}}\frac{h}{\tau_{e-g}}(1-f) - \frac{\Gamma_d}{A_d}a(2f-1)\frac{1}{N_{gsm}}\frac{S(t)}{\hbar\omega}, \tag{3}$$

where w, h and f represent the occupation probability of the wetting layer, the QD excited state, and ground state, respectively; N_{wm}, N_{esm}, and N_{gsm} are the maximum densities of carriers in each state; the spontaneous radiation lifetime of each state is denoted by τ_{ar} ("a" being "w", "es" or "gs"); τ_{a-b} denotes the relaxation time between any state "a" and state "b"; Γ_d is the active layer confinement factor; I is the injected current; a is the differential gain; V is the volume of the active layer; A_d is the effective cross-sectional area of the active layer; κ is the TPA coefficient and S(t) is the total input light power. The TPA generated carriers are taken into account by the last term in Eq. (1).

The gain of QD-SOA including nonlinear process such as carrier heating (CH) and spectral hole burning (SHB) effects is expressed as [27-30]:

$$g(t) = \frac{a(N - N_t)}{1 + (\varepsilon_{CH} + \varepsilon_{SHB})S(t)}, \tag{4}$$

where N and N_t are the GS carrier density, the transparent GS carrier density respectively; ε denotes the gain suppression factors. The refractive index of the active region is affected by the injected light and the change of temperature due to carrier heating. As a result, it will cause a phase change to any probe wave injected into the QD-SOA [13]:

$$\varphi(t) = -\frac{1}{2}\left(\alpha G_L(t) + \alpha_{CH}\Delta G_{CH}(t)\right), \tag{5}$$

where $G_L(t)$ is the linear gain factor of the device given by $e^{g(t)L}$, L being the effective length of the active layer; α is the linewidth enhancement factor of the device corresponding to band-to-band transition and α_{CH} is the linewidth enhancement factor of the device related to carrier heating process [31, 32]. The linewidth enhancement factor due to spectral hole burning is ~0 [10].

As noted previously, we use a QD-SOA-MZI to realize high speed all optical XOR and AND operation. The output of this scheme from the combination of two data streams can be expressed as:

$$P_{out} = \frac{P_{cb}(t)}{4}\left[G_1(t) + G_2(t) + 2\sqrt{G_1(t)G_2(t)}\cos(\varphi_1(t) - \varphi_2(t) + \varphi_0)\right], \tag{6}$$

where P_{cb} is the light power of the input clock signal; $G_1(t)$ and $G_2(t)$ are the calculated total linear gain factors. The primary parameters used in this simulation are presented in Table 1.

Table 1. Parameters used in the simulation.

Parameter	Description	Value
τ_{wr}	Recombination lifetime of WL	0.2 ns [26]
τ_{esr}	Recombination lifetime of ES	0.2 ns [26]
τ_{gsr}	Recombination lifetime of GS	0.1 ns [26]
Γ_d	Active QD region confinement factor	0.1
a	Differential gain	8.6×10^{-15} cm^2 [19]
$\tau_{w\text{-}e}$	Relaxation lifetime from WL to ES	3 ps [3]
$\tau_{g\text{-}e}$	Relaxation lifetime from GS to ES	10 ps [3]
κ	TPA coefficient	70 cm/GW [20]
L	Active region length	1.0 mm
α	Linear linewidth enhancement factor	4
α_{CH}	CH linewidth enhancement factor	1.2 [33]
ε_{CH}	Gain suppression factor of CH	0.3×10^{-17} cm^3 [31]
ε_{SHB}	Gain suppression factor of SHB	7.5×10^{-17} cm^3 [31]

4. Numerical Simulation and Discussion

In our simulation, a PRBS signal is used as input data and we assume the data stream pulses to be Gaussian pulses. The single pulse energy is 0.5 pJ and full width at half maximum (FWHM) is 1ps. The injected current to the QD-SOA is 250 mA. Figure 4 and Fig. 5 show the results of XOR performance without and with the consideration of TPA, respectively. Here we use the pseudo-eye-diagrams to show the output quality. The calculated "1"s and "0"s are superimposed on each other to plot the eye diagram. The output quality can also be quantitively characterized by the Q factor, which is defined as Q = $(P_1-P_0)/(\sigma_1+\sigma_0)$ [21]. Here P_1 is the average peak power of output signal "1"s and σ_1 is the standard deviation of all "1"s. P_0 and σ_0 are defined analogously for output "0"s. The output bit error rate (BER) is related with the Q factor by: BER ≈ $(2\pi)^{-1/2}\exp(-Q^2/2)/Q$ [8]. The primary reason for noise in this calculation is pattern effects resulting from the long recovery time of gain and gain-induced phase change. As we can see from Fig. 4 and Fig. 5, with TPA, the XOR output has a clearer eye diagram and thus a higher Q factor compared to the case without TPA. This is because that the TPA induced carrier pumping leads to a change in carrier recovery dynamics in QDs, which reduces the gain recovery time. Thus, pattern effects are reduced and a higher Q factor is achieved.

All-optical AND gates are realized with the same scheme of XOR gates as described in Section 2. The performance of AND gates is showed in Fig. 6 and Fig. 7. It also indicates the TPA in quantum dots can improve the output quality of the AND gates.

To furtherly investigate the effects of TPA on optical logic operations, we compare the dependence of Q factor of XOR gates on various parameters for these two cases. The calculated Q factor shows significant dependence on the injected current, initial pulse width, pulse energy and data rate as suggested in Fig. 8. As we can see in Fig. 8(a), by fixing the inject current at 250 mA and FWHM 1ps, the output quality degrades when one

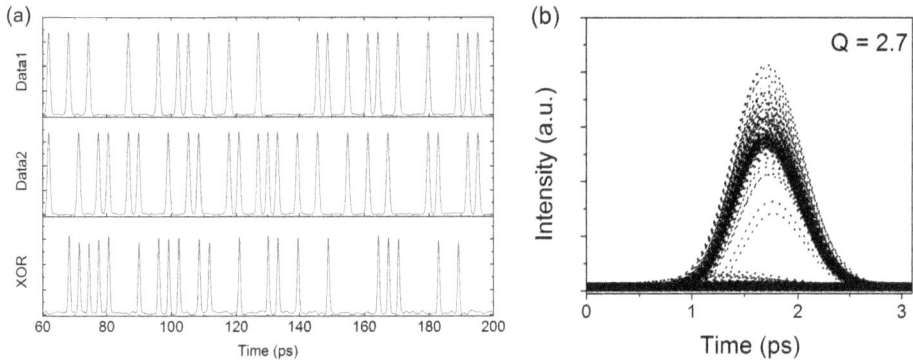

Fig. 4. Simulation results of XOR gates operating at 320 Gb/s without the consideration of TPA. The input single pulse energy is 0.5 pJ and pulse width is 1 ps.

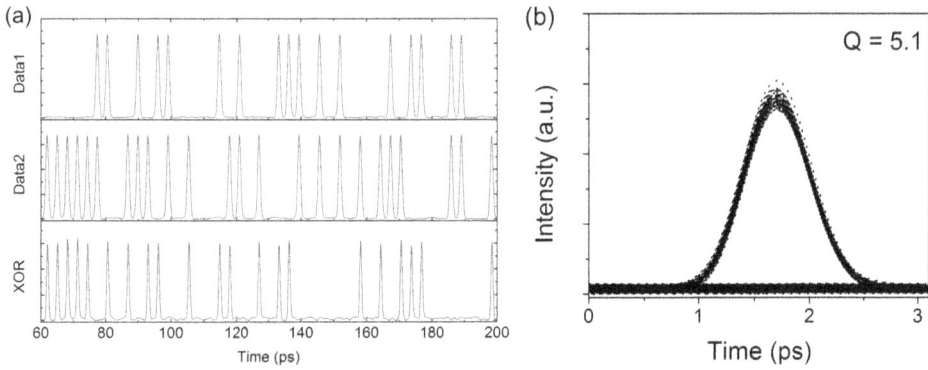

Fig. 5. Simulation results of XOR gates operating at 320 Gb/s with the effects of TPA. The input single pulse energy is 0.5 pJ and pulse width is 1 ps.

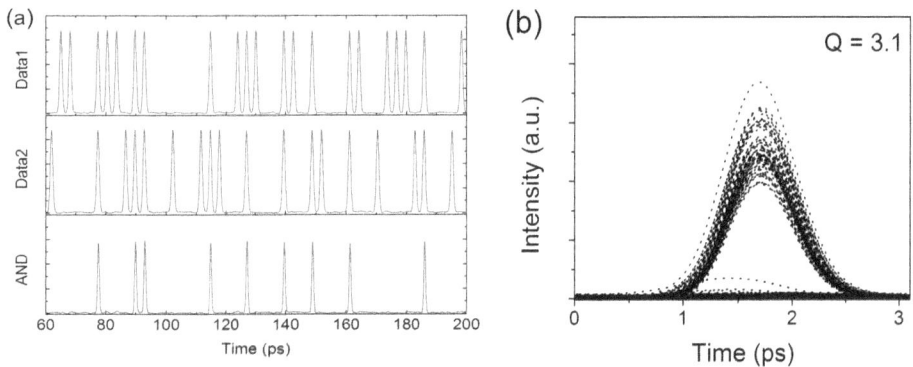

Fig. 6. Simulation results of AND gates operating at 320 Gb/s with the effects of TPA. The input single pulse energy is 0.5 pJ and pulse width is 1 ps.

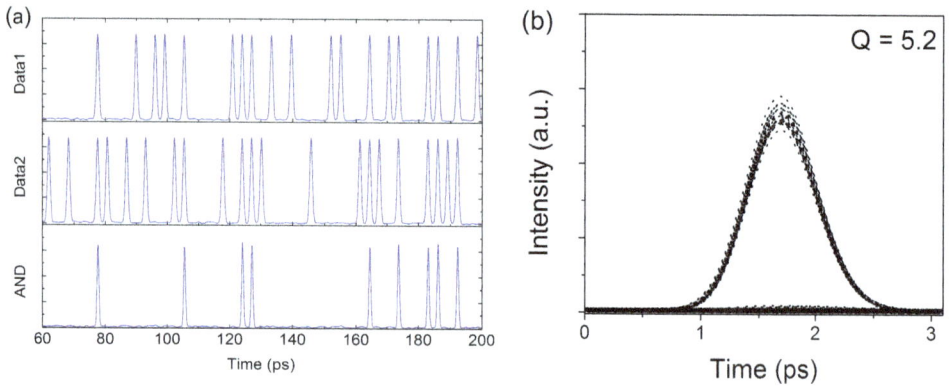

Fig. 7. Simulation results of AND gates operating at 320 Gb/s with the effects of TPA. The input single pulse energy is 0.5 pJ and pulse width is 1 ps.

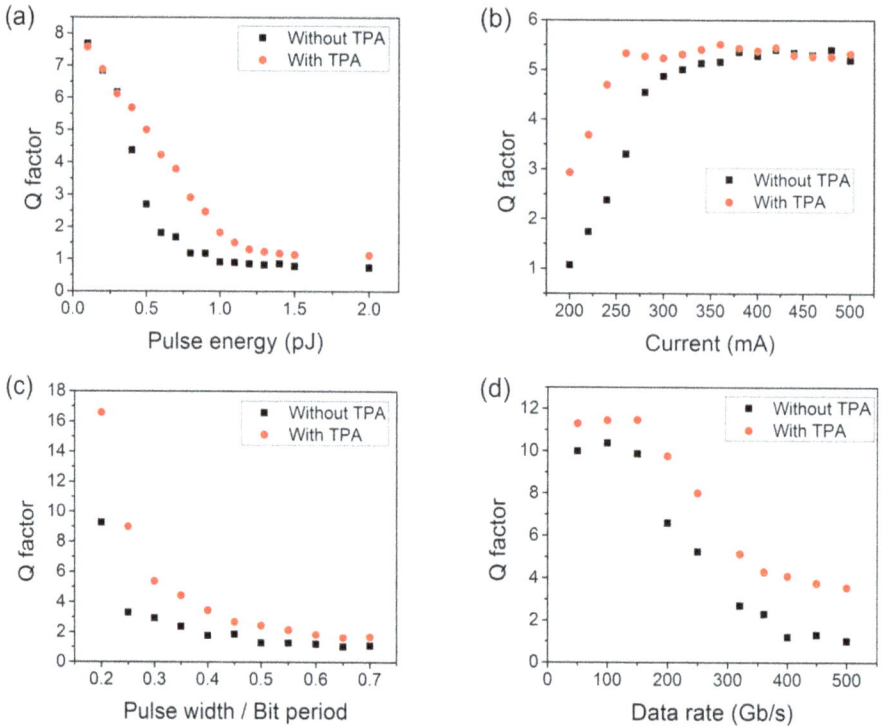

Fig. 8. (a) The dependence of Q factor on single pulse energy. The injected current is fixed at 250 mA and pulse width is 1 ps. (b) The dependence of Q factor on injected current. The input single pulse energy is fixed at 0.5 pJ and pulse width is 1 ps. (c) The dependence of Q factor on pulse width/bit period ratio at 320 Gb/s. The injected current is fixed at 250 mA and single pulse energy is set at 0.5 pJ. (d) The dependence of Q factor on data rate. The single pulse energy is 0.5 pJ and injected current is 250 mA. The ratio of the pulse width to bit period is fixed.

increases the single pulse energy of the input data for both with and without TPA cases. Without the effects of TPA, the Q factor drops very fast when pulse energy is increased from 0.1 pJ to 0.5 pJ; while with TPA taken into account, the decrease of Q factor becomes slower. This is easy to understand. As we increase the single pulse energy, the peak power of the pulses is increased. The TPA process is enhanced when the injected pulse train has a higher peak power, which will mitigate the depletion of carrier density of the active region of the device. As a result, the dropping rate of output Q factor will become lower. In Fig. 8(b), Q factor increases with the injected current until the current reaches a certain level (~275 mA with TPA and 350 mA without TPA) and then it saturates. With TPA, it saturates at a smaller current level compared with that without TPA. The inject current produces carriers to the wetting layer, thus in each energy level in the quantum dot the carrier density is able to recover to its initial state. The more carriers produced, the faster the recovery time will become. This can reduce the pattern effect due to long recovery time of the logic operation. The TPA process also contributes to the carrier recovery thus the Q factor can saturate at a lower current level than that without TPA. The Q factor decreases with the increase of pulse width and data rate for both cases. The overall Q factor of the TPA case is always higher than that without TPA. This demonstrates that TPA plays an important role in QD gain bleaching and reducing pattern effects for ultrafast optical logic operation.

5. Conclusion

We have studied the TPA effects of on all-optical logical operation based on QD-SOA MZI using the rate equation model. TPA contributes to the carrier dynamics by generating carriers captured into QDs on ultrafast timescale. This additional carrier pumping can improve gain recovery time in the QDs and thus reduce pattern effect in the all-optical logic operation. Our simulation shows the output has a higher Q factor considering the TPA effects. This QD-SOA based MZI scheme is suitable for all-optical logic operation at 320 Gb/s.

References

1. D. Hillerkuss, R. Schmogrow, T. Schellinger, M. Jordan, M. Winter, G. Huber, T. Vallaitis, R. Bonk, P. Kleinow, and F. Frey, "26 Tbit s-1 line-rate super-channel transmission utilizing all-optical fast Fourier transform processing," Nature Photonics **5**, 364-371 (2011).
2. H. Dorren, M. Hill, Y. Liu, N. Calabretta, A. Srivatsa, F. Huijskens, H. De Waardt, and G. Khoe, "Optical packet switching and buffering by using all-optical signal processing methods," Journal of Lightwave Technology **21**, 2 (2003).
3. W. Li, H. Hu, and N. K. Dutta, "High speed all-optical encryption and decryption using quantum dot semiconductor optical amplifiers," Journal of Modern Optics **60**, 1741-1749 (2013).
4. J. Bao, J. Xiao, L. Fan, X. Li, Y. Hai, T. Zhang, and C. Yang, "All-optical NOR and NAND gates based on photonic crystal ring resonator," Optics Communications **329**, 109-112 (2014).
5. H. Hu, X. Zhang, and S. Zhao, "High-speed all-optical logic gate using QD-SOA and its application," Cogent Physics **4**, 1388156 (2017).

6. K. Chan, C.-K. Chan, L. K. Chen, and F. Tong, "Demonstration of 20-Gb/s all-optical XOR gate by four-wave mixing in semiconductor optical amplifier with RZ-DPSK modulated inputs," IEEE Photonics Technology Letters **16**, 897-899 (2004).
7. Y. Fu, X. Hu, and Q. Gong, "Silicon photonic crystal all-optical logic gates," Physics Letters A **377**, 329-333 (2013).
8. S. Ma, Z. Chen, H. Sun, and N. K. Dutta, "High speed all optical logic gates based on quantum dot semiconductor optical amplifiers," Optics express **18**, 6417-6422 (2010).
9. H. Sun, Q. Wang, H. Dong, and N. Dutta, "XOR performance of a quantum dot semiconductor optical amplifier based Mach-Zehnder interferometer," Optics Express **13**, 1892-1899 (2005).
10. W. Li, S. Ma, H. Hu, and N. K. Dutta, "All optical latches using quantum-dot semiconductor optical amplifier," Optics Communications **285**, 5138-5143 (2012).
11. X. Zhang, W. Li, H. Hu, and N. K. Dutta, "Two-photon absorption-based optical logic," in *SPIE Defense + Security*, (International Society for Optics and Photonics, 2015), 946731-946731-946710.
12. S. Hendrickson, C. Weiler, R. Camacho, P. Rakich, A. Young, M. Shaw, T. Pittman, J. Franson, and B. Jacobs, "All-optical-switching demonstration using two-photon absorption and the Zeno effect," Physical Review A **87**, 023808 (2013).
13. W. Li, H. Hu, X. Zhang, and N. K. Dutta, "High Speed All Optical Logic Gates Using Binary Phase Shift Keyed Signal Based On QD-SOA," International Journal of High Speed Electronics and Systems **24**, 1550005 (2015).
14. X. Zhang and N. K. Dutta, "Effects of two-photon absorption on all optical logic operation based on quantum-dot semiconductor optical amplifiers," Journal of Modern Optics **65**, 166-173 (2018).
15. D. Leaird, S. Shen, A. Weiner, A. Sugita, S. Kamei, M. Ishii, and K. Okamoto, "Generation of high-repetition-rate WDM pulse trains from an arrayed-waveguide grating," IEEE Photonics Technology Letters **13**, 221-223 (2001).
16. X. Zhang, H. Hu, W. Li, and N. K. Dutta, "High-repetition-rate ultrashort pulsed fiber ring laser using hybrid mode locking," Applied Optics **55**, 7885-7891 (2016).
17. W. Li, H. Hu, X. Zhang, S. Zhao, K. Fu, and N. K. Dutta, "High-speed ultrashort pulse fiber ring laser using charcoal nanoparticles," Applied Optics **55**, 2149-2154 (2016).
18. H. Hu, X. Zhang, W. Li, and N. K. Dutta, "Hybrid mode-locked fiber ring laser using graphene and charcoal nanoparticles as saturable absorbers," in *SPIE Defense + Security*, (International Society for Optics and Photonics, 2016), 983630-983630-983638.
19. T. W. Berg, S. Bischoff, I. Magnusdottir, and J. Mork, "Ultrafast gain recovery and modulation limitations in self-assembled quantum-dot devices," IEEE Photonics Technology Letters **13**, 541-543 (2001).
20. H. Ju, A. Uskov, R. Nötzel, Z. Li, J. M. Vázquez, D. Lenstra, G. Khoe, and H. Dorren, "Effects of two-photon absorption on carrier dynamics in quantum-dot optical amplifiers," Applied Physics B **82**, 615-620 (2006).
21. N. K. Dutta and Q. Wang, *Semiconductor optical amplifiers* (World Scientific, 2006).
22. X. Zhang, W. Li, H. Hu, and N. K. Dutta, "High-Speed All-Optical Encryption and Decryption Based on Two-Photon Absorption in Semiconductor Optical Amplifiers," Journal of Optical Communications and Networking **7**, 276-285 (2015).
23. T. Akiyama, M. Sugawara, and Y. Arakawa, "Quantum-dot semiconductor optical amplifiers," Proceedings of the IEEE **95**, 1757-1766 (2007).

24. M. Sugawara, H. Ebe, N. Hatori, M. Ishida, Y. Arakawa, T. Akiyama, K. Otsubo, and Y. Nakata, "Theory of optical signal amplification and processing by quantum-dot semiconductor optical amplifiers," Physical Review B **69**, 235332 (2004).

25. P. Ridha, L. Li, M. Rossetti, G. Patriarche, and A. Fiore, "Polarization dependence of electroluminescence from closely-stacked and columnar quantum dots," Optical and Quantum Electronics **40**, 239-248 (2008).

26. S. Ma, H. Sun, Z. Chen, and N. Dutta, "High speed all-optical PRBS generation based on quantum-dot semiconductor optical amplifiers," Optics express **17**, 18469-18477 (2009).

27. T. Akiyama, H. Kuwatsuka, T. Simoyama, Y. Nakata, K. Mukai, M. Sugawara, O. Wada, and H. Ishikawa, "Application of spectral-hole burning in the inhomogeneously broadened gain of self-assembled quantum dots to a multiwavelength-channel nonlinear optical device," IEEE Photonics Technology Letters **12**, 1301-1303 (2000).

28. H. Hu, X. Zhang, W. Li, and N. K. Dutta, "Simulation of octave spanning mid-infrared supercontinuum generation in dispersion-varying planar waveguides," Applied Optics **54**, 3448-3454 (2015).

29. P. Borri, W. Langbein, J. M. Hvam, F. Heinrichsdorff, M. H. Mao, and D. Bimberg, "Spectral Hole-Burning and Carrier-Heating Dynamics in Quantum-Dot Amplifiers: Comparison with Bulk Amplifiers," Physica Status Solidi (b) **224**, 419-423 (2001).

30. X. Zhang, H. Hu, W. Li, and N. K. Dutta, "Mid-infrared supercontinuum generation in tapered As2S3 chalcogenide planar waveguide," Journal of Modern Optics **63**, 1965-1971 (2016).

31. O. Qasaimeh, "Linewidth enhancement factor of quantumdot lasers," Optical and Quantum Electronics **37**, 495-507 (2005).

32. J. Vazquez, H. Nilsson, J.-Z. Zhang, and I. Galbraith, "Linewidth enhancement factor of quantum-dot optical amplifiers," IEEE Journal of Quantum Electronics **42**, 986-993 (2006).

33. T. Newell, D. Bossert, A. Stintz, B. Fuchs, K. Malloy, and L. Lester, "Gain and linewidth enhancement factor in InAs quantum-dot laser diodes," IEEE Photonics Technology Letters **11**, 1527-1529 (1999).

www.ingramcontent.com/pod-product-compliance
Lightning Source LLC
Chambersburg PA
CBHW081519190326
41458CB00015B/5404